Microbes at war

From the Dark Ages to Modern Times

Jean Freney – François Renaud

In the memory of Louis Prillard from Chouzelot, in Doubs, France, died 11 October 1915 at the age of 20 years, from complications due to an infection in a shrapnel wound at the battle of Souchez, Pas-de-Calais, France.

Preface

From this highly informative book, rich in historical examples and anecdotes, one realizes that an endless war is still raging between microbes and men. The stories recounted here remind us of the past eras when both, the rich and poor paid heavily for infectious diseases and consequently had a much shorter life expectancy than us.

Scores of epidemics have left their mark on world history and often altered its course. Since millions of years, innumerable, ubiquitous, microscopic life forms have lived alongside and at times attacked human, animal and plant life. In the late 1960s, with the advent of efficient methods of diagnosis, development of protective sera and vaccines, therapeutic antibiotics, antivirals, antiparasitics, insecticides, etc., one believed that the war against microbes would fade out or even disappear. However, there was but a short truce. Since then, a good thirty-odd infectious agents have emerged or re-emerged and spread to occupy new geographical niches, for example, agents of legionellosis and mad cow disease, or viruses like Ebola, HIV, SARS, Nipah, Hendra, Chicungunya, West Nile, dengue fever, bird flu etc. Currently about one-third of human deaths are due infectious diseases.

While man is often a victim of these diseases, particularly in the Southern hemisphere, he is also largely responsible for their spread. This is due to several reasons: changes in human life-style and social behavior, globalization and free trade, lack of adequate hygiene and epidemiological surveillance in some countries, as well as global warming that promotes spread of diseases and arthropod vectors. Moreover, man is also responsible for the dangerous use of microbial pathogenic agents in biological warfare or terrorism.

The fight is on. This is what the two authors, Jean Freney and François Renaud, want to remind us of in "Microbes at War". May this book find a wide audience among the well-informed as well as the simply curious.

Alain Mérieux

Contents

Emergence of infectious diseases

Around two million years ago, prehistoric man learnt to make fire and assemble primitive tools from pieces of flint and eventually, to hunt for food. Complex proteins were introduced into the human diet through meat, and man became bigger, stronger and faster. However, as seen from fossil records, there was a price to pay for this progress. It was through this frequent contact with animals and meat products that zoonoses (diseases transmitted from animals to man) emerged, with the first cases of encephalitis, an infection in the brain and of the equally deadly, trichinosis. Then, around 10,000 BC during the Neolithic Age, man invented agriculture. This was primarily in the Fertile Crescent of the Middle East, covering countries such as Egypt and Mesopotamia. The agricultural revolution provided a better quality control and management of human diet. Instead of foraging in the wild, man learnt to cultivate land and grow plants on a large scale. The price paid for this progress was even higher. The first incidences of infectious diseases occurred in this era. Microbes and insects that breed in stagnant water in the irrigated lands and in the manure used as fertilizers, found their way into the water supply, which led to the spread of diseases such as cholera (a gastrointestinal infection transmitted through drinking water) and malaria (transmitted by mosquitoes). Urban environments in turn, became the focus of epidemics such as smallpox, measles, typhoid and scarlet fever.

Between 9000 and 3000 BC, along with the agricultural revolution, began the domestication of animals. In Eurasia alone, one can count up to twenty-two different domesticated animals such as sheep, dogs, goats, pigs, cattle, horses and camels. During the same era, other animals such as alpacas, llamas and guinea pigs were domesticated in America. As a result of this on-going contact between man and animals, a large number of germs were able to cross the species barrier and to adapt themselves to humans. Among these, small pox, for example, appeared in humans with the domestication of camels in Asia Minor. It is believed that the buffalo was the purveyor of leprosy in India and that tuberculosis likewise, was passed on to man from domesticated cattle. In South America, the guinea pig likely played a role in the transmission of Chagas disease, an infection caused by the parasite, *Trypanosoma cruzi*, and transmitted by an insect. Even today, this disease is rampant in the American continent, with as many as 90 million people being exposed to the parasite.

Age of demons, witches and deathly fumes

Before the advent of the Modern era, people were terrified of epidemics as they had no idea what caused them. It was an era of "miasmas", the demonical fumes or poisonous gases emanating from swamps and other "haunted places", and of punishments from God. The ancient Persians have described 99,999 demons that caused diseases! In some of their writings there is a description of "Al", the demon of scarlet fever: "Have you seen Al? It looks like a young girl with red cheeks and a purple body in flames".

Saturn devouring one of its children, 1815 painting by Goya, (Prado Museum, Madrid).

The demons being supernatural creatures, it was pointless to reason with them and impossible to punish them. The only option left for the doctors was to scare, bribe, or threaten them. In contrast to demons, the wizards and witches were human. Throughout the Middle Ages, old women who practised the art of medicine were accused of witchcraft and during epidemics were hunted down and exterminated by people who did not know better.

Interestingly, despite the complete lack of knowledge regarding the underlying cause of epidemics such as the plague, the notion of a contagion was conceived quite early. Since Classical Antiquity there are references to the initial use of the contagion as a "miasma", for hostile purposes. During this era, army chiefs who would well know where a fever was reigning, would strategically drive their enemies into this hostile land, and leave them there to die. At the end of the 19th century, and notably throughout the 20th century, with advances in microbiology and a deeper understanding of infectious diseases, microbes could be manipulated in a highly sophisticated manner, to the extent of being exploited as biological weapons, an idea that is a cause for concern even today.

1 "Plagues" of the Dark Ages

For a long period of time, several epidemics, or "plagues" as they were called, have struck people or armies in the field, and were the cause of their defeats or victories. Today, it is very difficult to know the origin of the word "plague", often used in the ancient times to mean scourge, because it did not have the same meaning as that acquired later with the Great Plague of the Middle Ages.

Trumpets of Jericho

The Old Testament talks about the fall of Jericho in the late 13th century BC during the conquest of the Promised Land by Joshua, the successor of Moses. The city of Jericho, seemingly impregnable, blocked the path of the Hebrews. Following directions from God, Joshua marched his troops around the city in a procession with seven priests in the forefront blowing the

Battle of Jericho, Julius Schorr von Carolsfeld (1794 – 1872).

trumpet with ram's horns, followed by soldiers carrying the Ark of the Covenant and the big troops making up the rear.

The procession marched around the city day after day for six days. On the seventh day while they were on their seventh round of Jericho, which would be the last, Joshua turned to the Hebrews and said "Shout". They all raised a war cry in unison and down came the walls. The entire population, including animals, was massacred, except for Rahab, a prostitute and her family who had given refuge to two Jewish spies. After looting the city of its riches and reducing it to ashes, Joshua threw his curse "Cursed be the one who rebuilds Jericho".

A scientific explanation may be given today, based on certain evidences that indicate the presence of an infection called schistosomiasis among residents of Jericho. Calcified fossils of the shellfish that transmit this disease were found in the dried wells of Jericho. The waters of Jericho being infested with these molluscs, the residents must have been debilitated by schistosomiasis and too incapacitated to return the attack by Joshua's army.

This infection may also be implicated in the fall of the Egyptian empire. In the Fertile Crescent of the Kingdoms of Sumer, Assyria and Babylon, ancient civilizations of farmers had developed and perfected techniques for intensive irrigation. The Egyptians used the flooding waters of the Nile for irrigating their farms. A regular contact with infested waters had

the same consequences among these people as among the people of Jericho. Around 660 BC, Egypt was subjected to various political conflicts and was finding it hard to resist attacks from the neighboring Assyrians, like in 525 from the Persians. The Egyptians, who had earlier withstood all attackers for 27 centuries, were now so weakened by chronic infections that they were unable to cope with the new challenges.

Many years later, soldiers from Napoleon's army suffered these same infections during their expedition in Egypt in 1799 – 1801. Dominique Larrey, the military doctor, had noted blood in the urine of a large number of soldiers from the troops and had attributed this to long marches in the blistering heat.

The baron Dominique Jean Larrey, by Girodet-Trioson, (Louvre Museum, Paris).

The "plague" of Philistines (1050 BC)

The belief that divine retribution arrives in the form of plagues after episodes of looting of temples and sacred places dates back to the time before the Ark of Covenant. In the 12th century BC, the Philistines, originally from Crete, sought refuge in Canaan (in the Gaza strip today). Around 1050 BC, during the battle of Aphec, they captured the Ark of Covenant, a wooden chest containing the tablets of Law, handed down by Moses to the Jewish people. The Ark was supposed to guarantee a victory to the armies that possessed it. Unfortunately, wherever the Philistines carried it, such as to Ashdod, Gath or Ekar, a strange disease characterized by "tumors in the private parts of the body" appeared. This was the first recorded outbreak of an epidemic in history (Bible 1, Samuel 5 and 6) called "plague of the Philistines". Faced with so much death and desolation, the Philistinians returned the Ark to their legendary enemies, the Hebrews.

This episode demonstrates that even in these ancient times, the concept of the contagion had already taken birth. The Testament of Solomon predicted that if the Temple of Jerusalem is destroyed by the Chaldeans, then the spirit of the plague would be released. This happened in 587 BC with the attack on the Temple by the cruel king of the Chaldeans, King Nabuchodonosor, during which the Ark disappeared. Later, when the looting ensued, the Chaldeans discovered copper containers and took them to be treasures. However, on opening them the demons were released and the "plague" struck again.

Much later, an almost identical scenario was replayed in the same geographical region when soldiers were looting a Greek temple in Babylon. The terrible "plague" of 165 – 180 AD travelled from this city to hit the Middle East and the Mediterranean, and even reached Rome, Gaul and Germany. Galen has described the symptoms of this epidemic in sufficient details so that it can now be identified as small pox.

The Ark of Covenant brought to the Temple (Conde Museum, Chantilly).

The Temple of Solomon was rebuilt in the 5th century BC. In 1945 AD, at Nag Hammadi in Egypt were discovered the first parchments written by the Christians in 400 AD, which relate other legends about the Temple. During the siege of Jerusalem in 70 AD by the future roman emperor Titus, the Temple was destroyed again and the ancient jars were discovered and opened by the Romans. The demon of the "plague" that had lain trapped in the foundations of the Temple since the time of Solomon was thus liberated. Suetonius, the Latin bibliographer of Titus recounts "the reign of Titus was marked by appalling disasters" and particularly by "a plague, the likes of which had never been seen before".

The siege of Jerusalem (701 BC)

The pharaoh, lulled by his visions of grandeur had treated his army with contempt, in the belief that he will never need it. Around 700 AD, when Sennacherib, King of the Assyrians, began to march across the borders of Egypt, the troops of the Pharaoh refused to fight for him. "It was a disaster" writes Herodotus (around 482 – 425 BC).

Sennacherib

The plague of Ashdod, by Nicolas Poussin (Louvre Museum, Paris).

The Assyrian army set up camp at Pelusium, a flat, salty region, northeast of the Egyptian border. The Pharaoh, who was also a priest of God Ptah was greatly distressed. Regretting his pride, lamenting and weeping he fell asleep, when God appeared in his dreams. Ptah advised him to forget about his soldiers and call instead all the merchants, craftsmen and other people to raise an army that would meet Sennacherib. God himself promised to help him secure victory.

With great confidence the Pharaoh marched at the head of his peculiar army to face the enemy. In the dead of the night, nothing moved, except for the thousands of field mice that crept into the Assyrian camp and started gnawing at their leather quivers, strips of the shields and bowstrings. The next morning the Assyrians were helpless and unarmed as all their weapons had been rendered useless. In the Dark Ages, rodents destroying the trappings of an army was seen as a bad omen, signaling an immediate disaster. Since long, the appearance of hordes of rodents was noted as a harbinger of epidemics. The Assyrians, now scared, lifted the camp and fled. The Egyptian army made up of riff raff was thus able to inflict heavy losses on that of Sennacherib. Herodotus claimed that he heard the story from the lips of the priests of God Ptah, who showed him a statue of the Pharaoh holding a rat. Most historians believe that there is some truth behind the story.

The Old Hebrew Testament, talks about the sudden defeat of Sennacherib in 700 BC, but claims that it happened at the gates of

Jerusalem: "an angel sent by God smote 175,000 Assyrian soldiers".

Josephus, a Jewish historian, adds to the account given by Herodotus in his writings from 93 AD. The bad omen of rats destroying leather was not the only reason for the shameful retreat of the Assyrians. According to Josephus, Sennacherib had been warned that a formidable Ethiopian army would come to the rescue of the Egyptians. Further, citing Berossus, a historian from Babylon (300 BC), Josephus confirms that an epidemic (plague or typhus) had killed 185,000 Assyrians in their flight from Egypt through Palestine. In 1938, in Lachish, a town near Jerusalem, the English archaeologist, James Lesley Starkey discovered a cave filled with charred and broken bones of about 1500 people, who must have been abandoned by the fleeing Assyrians. It was a mass grave with 1000 arrowheads made of iron. Curiously, pig bones were attached to the human bones. The Jews considered the pig as an unclean animal, which would indicate that these bones were indeed those of the soldiers from Sennacherib's army.

An angel destroying the army of Sennacherib, by Gustave Dore.

The defeat of the Assyrians could also have been due to malaria. They had left the torrid lower plains of Jordan, situated below sea-level and ventured into the Palestine plateau all the way to Jerusalem, at a height of 790 meters. This migration from the torrid plains to Jerusalem where the weather was colder and climatic conditions harsh, favored the development of the protozoan, malarial parasite. Similarly, in 1917, malaria decimated half of the English soldiers who had been convalescing close to the same marshy areas of Jordan.

The "plague" of Athens (431 - 429 BC)

In the middle of the 5th century BC, Athens and Sparta, dominated the Greek world. In his "History of the Peloponnesian War" of 431 – 404 BC, Thucydides describes an epidemic that devastated Athens at the beginning of the summer of 431. In fact he contracted the disease himself, but survived. Within two years, the population was reduced by about a quarter or one-third. The powerful General, Pericles himself succumbed to it in 429.

According to Thucydides, the "plague" began in the second year of this War. The disease

Pericles, British Museum, London

broke out in Piraeus. Tens of thousands of inhabitants took refuge in Athens as it opened its doors to them on instructions from Pericles. Thus 400,000 people crowded inside the city walls. The first epidemic that lasted two years, was followed by a respite of a year and half, and when it resurfaced again it was lethal. According to Diodorus, 4,400 hoplites, 300 cavalry, a large number of soldiers as well as 10,000 women and slaves died in this "plague".

Undoubtedly, the considerable increase in population experienced by the city due to the entrenchment of an entire population behind the walls of Attica contributed to amplify the severity of the epidemic. Several possibilities have been put forth for what this mysterious disease could have been: smallpox, typhus, dengue fever, or fever due to staphylococcus toxin, mycotoxins, arboviruses, or Ebola virus, while plague appears to be the least likely. Another hypothesis is that of relapsing fever caused by *Borrelia recurrentis*, which is carried by the louse. Today, one of the largest endemic area for this disease is in Ethiopia. Thucydides writes, "the disease appeared for the first time in Ethiopia". This epidemic was clearly that of the louse-borne relapsing fever, which in the past had been associated with migrating populations, as during the Peloponnese War. However, the infectious etiology is not convincing. If indeed there were rats that carried the plague bacillus in Athens, why would they have infected others but spared the Spartans?

Read has put forth the hypothesis of an intoxication caused by an ergot that infests rye. This *"mal des ardents"*, "ergotism" or "fire of Saint Anthony", known since the 10th century is caused by ergot, a hallucinogenic, fungal parasite of cereals. In a matter of days it can affect an entire region. Several treatments have been available since 1070, the most well-known being those given by the monks of the Order of Saint Anthony.

The Temptation of St. Anthony, by Mathis Grünewald Gothart, Unterlinden Museum, Colmar
©BPK, Berlin, Dist RMN/Jochen Remmer.

Some of the noteworthy symptoms of this sickness depicted in paintings by Jerome Bosch and many others are: an internal burning sensation making the patient seek coolants, a good physical resistance despite serious clinical signs, insomnia and amputations. There is a striking resemblance between the clinical signs described by Thucydides and those of ergot poisonings. A simple explanation as to why the disease remained within the walls could be that the refugees besieged in Athens did not share their ergot-infested cereal with the Spartans.

Thus, we have here a horrific disease that is infectious and toxic, in the backdrop of ignorance and anxiety. On the resulting panic is grafted a fatalism conferred on the Athenians by the salvational deity, Demeter. Today, what would we call this pseudo-plague that killed Pericles but spared Thucydides? Was it an infectious disease or rather an infectious disease plus ergotism? Or a new, unknown disease? Two thousand five hundred years later, we still do not have any convincing proof of the etiology of this disease. Was it ergotism then? Quite possibly, given the uncanny clinical similarities between ergot poisoning and the "plague" of the Athens.

Death of Alexander the Great

In 323 BC, the victorious king, after having marched his army all the way to India, stopped over in Babylon. After an evening of heavy drinking in the company of Medius, he went to bed and was seized by a high fever. He died the evening of June13[th], less than a month before his 33rd birthday. The symptoms, such as increase in volume of the spleen, abdominal pains and delirium, indicate that Alexander was most likely suffering from a pernicious form of malarial infection, which was endemic in lower Mesopotomia. He had just returned with a few vessels from a trip in the marshes of Euphrates near Babylon. Some invoke typhoid fever, or acute pancreatitis or even an overdose of hellebore, which was used in the Dark Ages for its analgesic and cardiotonic properties.

Alexander the Great on his horse, Bucephale. Pompei, National Archeological Museum, Naples.

2 The first human interventions

The choice of battle grounds

The idea of defeating the enemy by spreading diseases, dates back to the Dark Ages. Indeed, well before the discovery of bacteria and viruses, some strategists had somehow exploited infectious phenomena in warfare. Epidemics have often played an important role in the outcome of battles.

The military chief and Greek writer, Xenophon (430/425 – 355/352 BC) recommends in his memoires to carefully monitor the health status of the soldiers and above all, to "establish campsites in safe areas, avoiding in particular, wetlands and places where water and air are unclean and could cause diseases".

Xenophon, Prado Museum, Madrid.

Brennus and the siege of Rome

Brennus, whose real Gallic name is Brennos, from the root Brenn, meaning "warlord", was a chieftain from Gaul in the 4th century BC. In 390 BC, after his troops had devastated Rome, he demanded a ransom in gold in exchange for the liberation of Rome and scornfully insulted the Romans with his famous words, "Vae Victis" or "Woe to the vanquished". And yet, as described by the historian Livy (Totus Livius, 1st century BC), the Gallic tribes, having camped in swampy areas, suffered from fevers and various diseases and had to flee.

The sieges of Syracuse

Biological warfare can also be indirect. For example, some strategists would compel the enemy to prolong their stay in unhealthy areas, as we will see in the following example.

Sparta versus Athens

Brennus placing his sword on balance. Paul Lehugeur, 19th century.

Alcibiades from Athens, a nephew of Pericles, was 30 years old when he decided to resume war against Sparta. Playing on the rivalry between the two Peloponnese cities, he rallied alongside the Athenians, and pitted them against the Spartans. In 418 BC, war broke out again. In 416 BC, Alcibiades persuaded the Assembly to resume political expansion in the Western regions and to attack Sicily. The Athenians dispatched a massive expeditionary force, commanded by Nicias and Alcibiades against Syracuse, the most powerful city in Sicily. Meanwhile, back in Athens there was a political turmoil because a large number of hermai placed in the town squares and parks were found mutilated. The hermai were stone pillars carrying the head of the God Hermes and were revered by the Athenians. Alcibiades was accused of the being responsible for the destruction of the hermai and a ship was dispatched to bring him back from the expedition and stand trial.

Alcibiades, Capitole Museum, Rome.

For its part, Syracuse had sent its emissaries to seek help from Sparta to counter a possible Athenian attack. The emissaries met with an unexpected advisor: Alcibiades, who had escaped from the boat that was taking him back to Athens. Defecting to the enemy camp, he assured the Spartans that there was a real threat of a takeover of Sicily by Athens. Sparta thus agreed to send its troops to help Syracuse. The Athenian generals, Lamachos and Nicias tasted some initial successes in the spring of 416. However, the Syracusian General, Hermocrates, organized a force of resistance and fortified the city. He strategically compelled Nicias to prolong his camp in a damp plain, known for the disease of the marshes, probably malaria. Depleted by diseases, the entire Athenian expedition had to lift the siege and retreat soon after. Hermocrates thus brought defeat to Athens, which never quite recovered from this disaster.

Siege of Syracuse, step by step

1 - Athenians approached Syracuse from the northwest (Euryalus) and after a brief exchange were ruling over the plateau of Epipolae. Then

they built two forts, one at Labdalum in the northern front and the other, a rounded one to control the southwest. From the latter, they raised two parallel walls to cordon off the enemy town.

2 - The Syracusans in their turn, erected a wall, which was quickly destroyed by the Athenians.

3- The Syracusans constructed a dike and a fence along the marshes, but this structure was also captured by the Athenians after coordinated attacks from the infantry and the naval fleet, who cut across the marshes on boards.

4 - The Athenians strengthened their wall on the southern front and on the side to protect the fleet. At this time Nicias, the Athenian commander, left the northern wall unfinished which proved to be a serious error in judgement.

5 - The Syracusans asked for help from the Spartans, who declined to send their troops, but sent their

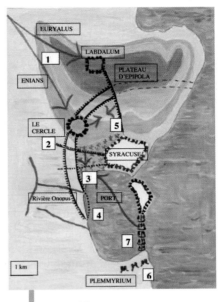

Siege of Syracuse
(by Mireille Chanteur.

general Gylippus instead. He managed to assemble 3000 troops and entered Syracuse without encountering any resistance. Gylippus seized control of Labdalum and built a wall between the city and the fort. This is the key point in the conflict.

6 - Suspecting defeat, Nicias asked his high command in Athens for permission to retreat but instead received reinforcements. He then put up three forts at Plemmyrium, south of the port and transferred his fleet over there.

7 - Both the adversaries were now reinforced. Gylippus attacked from the sea and land and after a series of exchanges, occupied Plemmyrium. He thus trapped the Athenian fleet inside the harbor and sealed the exit with a string of boats. In the meantime, the Athenian infantry was trapped in the marshes and eventually succumbed to high fevers. Ironically, these were the same sailors who had abandoned their boat in an attempt to escape by land. The Syracusan cavalry and light troops left them no respite. Weakened by fevers and thirst, those who survived, surrendered. This was the biggest defeat faced by Athens. Together with its allies, the Greeks, Estruscans and Italians, Athens lost 150 – 200 naval vessels, and 40,000 – 50,000 men.

Rome versus Carthage: "the Mediterranean turns into a Roman lake"

At the beginning of the 4th century BC, Carthage dominated the Mediterranean region. The Carthaginians had destroyed and taken possession of the suburbs of Syracuse, as well as the temples of Ceres and Proserpine. It was in this prosperous period that the disease first appeared among the Carthaginians and then spread so rapidly that the corpses abandoned without a proper burial soon made the place uninhabitable. The infected patients would die within 5-6 days in great pain. Diodore of Sicily has described this epidemic, attributing it to the vengeance of goddesses displeased by the destruction of their temples and secondly, to the hot weather and the crowding of thousands on men in low, swampy areas. According to his writings of 396, the Carthaginians who were decimated by the disease had been camping in the same region as the Athenians, about twenty years ago. Faced with considerable losses in the troops, the Carthaginians ended not only their siege but consequently their influence over Sicily.

This epidemic had a considerable impact on history because Sicily was a key point of control in the Mediterranean and having escaped from the Carthaginians control, turned into a "Roman lake". In fact, this event had taken place about a hundred years before the First Punic War between Rome and Carthage that was staged at Sicily. The Romans took 23 years to expel the Carthaginians who had conquered a part of the island, but only through huge sacrifices. It will not be far from the truth to say that if the siege of Syracuse of 416 BC had succeeded, the position of the Athenians in Sicily would have been unassailable and would have prevented the Romans from settling here for a long time to come.

Several years later, the two armies were to meet again at the same place, once again to be affected by a mysterious epidemic.

Siege of Astacos

In 364/363 BC, Clearchus, an aristocrat and a former pupil of Plato and Isocrates, established a particularly vicious tyranny that last 12 years in the city of Heraclea of Pontus, on the Black Sea (today, Karadeniz Eregli in Turkey). In an attempt to get rid of the dissident population of Heraclea, Clearchus enlisted men between 15-65 years old from this city to take part in the siege of the Thracian city of Astacos (today near the city of Izmit, west of Turkey) during the hot summer of 360. Astacos was surrounded by swamps. Pretending to "perform the most important part of the siege", Clearchus and his men occupied higher regions that had the advantage of tree cover and fresh air and water. Meanwhile, the men from Heraclea, made to camp down in the very hot swamp got only the stagnant water as nourishment. The siege lasted all summer. All the men from Heraclea died of fevers and Clearchus returned with his mercenaries, claiming that plague had hit the men in the swamp. A few years later, by a fitting turn of events, Clearchus was murdered.

Attila "the Scourge of God" and the Huns

Towards the end of the first century AD, a new race of fearsome warriors, the Huns, people from northern China emerged from the steppes of Mongolia and moved over to southeast Europe. Their exodus was likely

The Huns in a battle against the Alans: illustration (1873) by Johann Nepomuk Geiger (1805-1880).

provoked by poverty and famine. They appeared on the Volga in 370 AD and gradually made their way eastwards. Around 435, Attila successfully united all the tribes and established a formidable empire of the Huns situated in the great plains of what is today Hungary. Then they displaced the Germanic people, such as the Alans, Visigoths and Ostrogoths and came to occupy central Europe. The Huns brought with them new "plagues", unknown to the Europeans. Conversely, they also encountered diseases unknown to them, such as malaria. Around 451-454 AD, they reached Gaul and north of Italy. It is generally believed that the Pope Leo I who later became Saint Leon, had persuaded Attila to spare Rome. In fact, while the Huns were trying to attack Rome, they were apparently affected by a strange epidemic, quite likely malaria, which prevented them from seizing the imperial city. Recently, during an excavation at Lugano, scientists discovered traces of *Plasmodium falciparum*, the causative agent of malaria, dating back to 450 AD. These were found in samples of leg bones of a three year old child as well as in fifty other bones of children. This clearly demonstrates the nature of the malarial epidemic that raged in Italy at the time of Attila.

The meeting of Leo the Great (painted like the portrait of Leo X) and Attila: fresco by Raphael, 1514, Stanza della Segnatura, Palazzi Pontifici, Vatican.]

The first human interventions

Rome and the epidemic of plague

The Antonine plague broke out in Mesopotamia in 164 AD among the troops of Lucius Verus, adopted brother of Marc-Aurelius Antoninus, while camping on the eastern borders of the Empire. The epidemic was limited to the east for two years, ravaging the armies of Avidius Claudius, who had been sent as reinforcement to quell the revolt staged in Syria. The infection spread to Rome in 166 with the return of the troops, and then to Gaul and the banks of the Rhine and eventually to all parts of the world, causing numerous casualties. This event corresponds to the first gap in the expansion of the Roman Empire, which until 161 continued to grow and push its frontiers. That year, a Germanic tribe forced through the north-eastern border of Italy. The resulting terror and chaos lasted 8 years. In 169 all the Roman forces joined together to fight the barbarians. But the latter seem to have been supressed rather by infections spread by the legions, as evidenced from the fact that many of them were found dead, without any signs of injuries on a battlefield. The epidemic continued until 180, claiming the death of Emperor Marc-Aurelius himself. He died seven days after contracting the disease, having declined to meet his son, for the fear of infecting him. After a brief respite, the epidemic returned again in 189.

Emperor Marc-Aurelius (Capitol Museum, Rome).

The second time round, the epidemic spread less than the first one, but more importantly, affected Rome. At its peak, it killed 2000 people per day. This "plague" of 164-169 is also called "Plague of Galen", who described it in detail. According to most historians, this is in fact the first description of small pox.

After 189, there was no mention of a "plague" this severe, until 250, the year of onset of the Great Plague of Cyprian, under the reign of Trajan Decius, which indisputably changed the course of history in Western Europe. The nature of the infection still remains unknown. Cyprian, the Christian bishop of Carthage, looked upon it as a divine vengeance and has described the symptoms of the infection as diarrhea, retching and vomiting, ulceration in the throat, high fever and gangrene in the hands and feet. The symptoms are reminiscent of ergotism, caused from eating rye bread infected by a fungus, *Claviceps*. The rapidity of the spread of the disease would suggest that it was like the "Spanish flu" of 1918-1920, or may have been an epidemic of typhus. The acute phase of the "Plague of Cyprian" lasted 16 years, creating a general feeling of helplessness and panic. Thousands of people left the countryside to crowd into cities, leaving behind immense tracts of uncultivated lands.

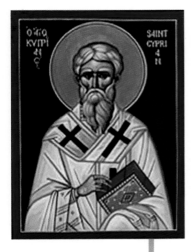

Saint Cyprian.

Despite the war in Mesopotamia on the Eastern border and in Gaul, the Roman Empire managed to overcome this catastrophe. However, in 275, the legions were driven from Transylvania and from the Black Forest to the Danube and the Rhine. The situation was serious enough that Emperor Aurelian decided to fortify the city of Rome itself.

Conquest of Damascus

When Caliph Omar decided to conquer Damascus, plague was reigning in the city. According one of the principles formulated by Hippocrates and later adopted by Islam, he maintained his troops in the desert, waiting for the epidemic to subside and then in 637, attacked the city to capture it without much difficulty.

Germanic expeditions in Italy

These provide a good example of how several armies were defeated due to an epidemic. For example in 963 and 964, the army of emperor Otton I was taken by surprise by an epidemic of the "plague", particularly fearsome as it caused death within 24 hours. Similarly, the army of Henri IV was hit during the fall of Rome in June 1083. The worst came in 1084 when the entire troop was destroyed. The same phenomenon happened again in 1137, in the army of Lothair.

Soon after the capture of Rome by the troops of Frederick Barbarossa in 1167, an epidemic of bubonic plague completely decimated the Emperor's army. The south of Italy in its turn was affected by the plague in the winter of 1190-1191 during the siege of Naples by Henry VI. The Emperor himself was affected and fled to Capua.

The Crusades

During the crusades, there was considerable mortality caused by infections, like during the capture of Antioch in 1097-1098. The "plague" was first seen in women and children who accompanied the Crusaders. It spread rapidly due to the general poor health resulting from lack of food and also due to the continuous rain that year. It is estimated that from September to November 1097 this disease claimed 100,000 people. The nature of the disease is not clear. Fresh reinforcements of 1500 Germanic soldiers who arrived, perished within days. In the summer of 1099, which was particularly hot, several dead bodies were abandoned without a proper burial and this favored the spread of the epidemic for another year.

During the second of the Crusades, a severe epidemic hit the French army of King Louis VII in Attali in Asia Minor. The epidemic spread rapidly and destroyed the entire population of the town. Similar events occurred again during the third Crusade after the death of Frederick Barbarossa in June 1190, during the siege of Antioch. From this Germanic army, only 5000 men of the infantry and 700 of the cavalry survived. The troops of Saladin were also affected, but much less severely, as they lost only 100 - 200 soldiers per day. During the siege of Saint John of Acre, which lasted from August 1189 to July 1191, duke Frederick of Swabia, succumbed to the disease on 20th January 1191.

During the crusade against the heretics, an epidemic broke out in Egypt among the Crusaders who were already suffering from dysentery that had appeared during the siege of Damietta in December 1217 following a period of unending, torrential rains. According to the chronicles, the affected people died horribly, suffering from intense pain in their feet and joints, their gums swollen, their teeth lost, their hips and arms

The capture of Antioch by the Crusaders, engraving by Sebastian Mamero, 1490.

blackened and putrefied. A sixth of the army of the pilgrims was destroyed by this disease which could have been a severe form of scurvy. The besieged troops also suffered the epidemic which was associated with another infection, the Egyptian endophthalmitis. According to the chroniclers, "a horrible scene awaited" the pilgrims when they took possession of Damietta. The streets and houses were strewn with cadavers and an appalling odor pervaded the town. Of the 80,000 that were housed in the city before the siege, only 3,000 survived and of these, only a hundred were uninfected.

The death of Saint Louis of Tunis during the 8th crusade in August 1270 will be recounted later.

3 The first use of biological weapons in history

The Scythians

Indo-Europeans, probably related to the Iranian and Slavic branches, the Scythians established themselves rather late (10th century BC) compared to the other Indo-Europeans. Their geographical domain was the steppes, grasslands and woodlands north of the Black Sea, extending up to the Volga in the East and up to Caucasus and the Danube in the South. For nearly a thousand years the Scythians occupied this immense steppe and established their original culture throughout the region. Essentially nomadic by nature, they needed to live in large areas. Between 700 to 550 BC, the centre of their culture was in the southeast, near Kuban and the Taman peninsula, extending up to southern Ukraine, south of Kiev. They have also left their mark in the north-east all the way into the region of Saratov. Although the Scythians did not have a script, the Indo-European languages are said to have originated from their language. The men are described as bon-vivant, lazy and heavy drinkers. They were polygamists and when they died, they were buried along with one of their wives and their horses.

They were known to be fierce warriors. They were the first people to ride a horse, although horses had been domesticated since a long time. This gave them a real advantage and afforded them their first victories in war, penetrating into Asia. The speed at which they could attack their enemy, from a vantage point, astride a horse was a key feature in their success. They followed some gory war customs: they would cut off the enemy's heads and drink from the skulls, just to terrorize others trying to resist them. The Persian King Darius mobilized about 700,000 men to quell the Scythian force, but had to beat a retreat. The Scythians never confronted the enemy face to face. They stripped everything off their land before the

A Scythian warrior on a horse.

enemy arrived, leaving behind neither the village, nor a single house, nothing for the enemy to plunder, except an endless steppe. The enemy would be fighting thin air! The Scythians were true artists as well, with a special liking for gold. One finds this precious metal indifferent objects such as tiaras, necklaces, rings, bracelets etc. and even on vessels and clothes and weapons.

The word Scythe means "archer". The Scythians employed various ways of making poison arrows. Their arrows were prepared with care, by plunging them in decomposing bodies or putrefying blood and "left at temperature of manure for maturation". The organic materials on the arrowheads decomposed and allowed the proliferation of pathogenic bacteria. The arrows would thus inoculate the wounded enemy with germs from cadavers, such as *Clostridium tetani*, the causative agent of tetanus or *Clostridium perfringens*, the bacteria causing gas gangrene. Death by infection was inevitable under these conditions.

Alternatively, they killed poisonous vipers that had just given birth, since these were easy to capture, and then with total apathy, would leave their bodies to decompose. The next step was performed by the Shamans, who were very important people in the Scythian culture, as they would supervise the preparation of venoms. Different ingredients would be mixed together. The first step was to mix human serum with the poison and smear it on the arrows. Another preparation consisted of mixing human serum mixed with animal excreta in leather pouches that were buried underground for putrefaction.

In a battle, each archer was capable of shooting about twenty arrows a minute. The enemy soldiers hiding behind the shield had surely heard about the terrible effects of the *scythicon* and the panic created by the poisoned arrows was undoubtedly intentional.

The same technique of dipping arrows and other weapons in the decomposing flesh of a viper was employed by the defenders of Harmatelia (probably Mansura in Pakistan) against the armies of Alexander the great in 326 BC. This choice was entirely appropriate, as it was shown recently, that apart from the pathogenic organisms present in the decomposing flesh, vipers retain a large amount of their fecal matter for several months in their bodies, which further increased the bacterial load. Much later, during the Vietnam war, the Vietcong used similar methods against American soldiers. These consisted of smearing excreta on sharp weapons, called 'pungee sticks", which would induce sepsis in deep wounds.

From the Dark Ages to Modern Times

Hannibal

In 241 BC, Hamilcar Barca commander of the Carthaginian army, lost the First Punic War against Rome. The peace treaty that ensued was humiliating for Carthage which had to lose a large part of its empire. Hannibal, son of Hamilcar Barca, was just over 9 years old. He was trained in the spirit of military campaigns by the Greek army generals and was influenced by the Hellenistic culture.

On the death of his father in 229 BC, he was placed under the orders of his stepbrother, Hasdrubal and together they consolidated the power of Carthage in Spain. After the death of his brother Hasdrubal in 221, he took over the army command. He made a plan of attack on Rome, not by the sea, which was the only imaginable route at that time,

Hannibal, National Archeological Museum, Naples.

but by land, going through Spain, Gaul and Northern Italy. In 219, he seized Sagunto, a Spanish town that was allied with Rome, and this triggered the Second Punic War. In 218, the Roman army decided to capitulate Carthage by creating two different fronts. One part of the army attacked Spain by the sea route, while the other was stationed in Sicily. Meanwhile, commanding an army of 38,000 men and 37 elephants, Hannibal was advancing across the Pyrenees, going up to an altitude of 2,000 meters above sea level, fighting against cold weather, snowstorms, landslides and also the hostile local tribes and marauders.

Passage through the Alps.

Then, he most likely took the route via Col de Petit Saint-Bernard, but still lost 15,000 men before reaching Turin in Italy.

The Roman army was forced to repatriate its troops based in Sicily, but faced a huge defeat in Cannae in 216, losing 45,000 men and about 20,000 prisoners from the ranks. This was a terrible disaster. Despite this victory, Hannibal did not take the city of Rome. He was stayed put at Cannae (now Barletta), about 400 km from Rome. Then in 202 BC, the Roman army led by Scipion, inflicted heavy losses on the Carthaginian army and Hannibal was defeated in Zama, today close to Maktar in Tunisia. A peace treaty was signed in 201 BC. In 196 BC, Hannibal was designated chief magistrate of the city of Carthage. He revamped the city but still remained a dangerous man for Rome and its allies and was exiled. He then joined the court of Antioch in Syria and helped the King in wars against Rome. However, after serious disagreements, the King of Antioch drove him out of his court. Hannibal then joined the King of Bithynia in Asia Minor. It was in 184 BC, during a naval battle against King Eumenes II of Pergamum, who was supported by Rome, that Hannibal first successfully experimented with a new form of biological warfare. His men threw clay pots filled with poisonous snakes on the enemy naval ships! The fighters, amused at first, would take fright at the sight of these "serpents from heaven". Having to fight two enemies, they would lose the naval battle. This tactic of using living organisms achieved the dual aim of scaring and fighting the enemy. Under pressure from the Romans, Prussia started to gradually withdraw its protection to Hannibal, and were almost at the point of handing him over to the Romans, when in 183, Hannibal committed suicide by taking poison from a ring that he always wore.

Contaminating wells

The contamination of wells and water reservoirs with human or animal cadavers was known since the Dark Ages and was used equally well by the Persians, the Greeks under Alexander the Great, and later by the Romans. In North Africa, during the Punic Wars (264-146 BC) against Carthage, the Romans allegedly threw dead animals in enemy waters.

Barbarossa

Frederick I of Hohenstaufen, known as Frederick Barbarossa, the Germanic emperor from 1152 to 1190, was called upon by the Pope Adrian IV to campaign against the cities of Lombard. He was crowned the Emperor in Rome by the Pope, after taking control of the city from Arnold of Brescia, a monk who was later hanged.

Emperor Frederick I Barbarossa, Vatican Library.

Tortona, in Piedmont, was at that time an ally of Guelph, and supported the Bavarian dynasty of Welfs. The latter claimed the throne of the Holy Roman Empire against the dynasty of Hohenstaufen. In 1155, in order to take control of this hostile city, Barbarossa strategically threw bodies of dead animals and humans in the wells and in addition burnt them by pouring tar and sulphur so that the water was unfit for consumption.

Iroquois

In 1710, Iroquois, the indigenous people of North America, contaminated drinking water supplies by throwing in animal skins, killing more than a thousand French soldiers.

The Southerners

Siege of Vicksburg – the 13th, 15th and 17th Corps of Army commanded by General Grant (Kurz and Allison, Congress Library.

Similar tactics were seen during the American Civil War when the Southern troops unsuccessfully attempted a bacteriological attack on the Yankees of the North. While retreating from Vicksburg in 1863, the confederation of General Johnston poisoned wells to slow down the advancing troops of General Sherman. The city being located in the swampy areas bordering Mississippi, there were a large number of casualties from malaria that devastated the region.

Sieges of citadels

The assailants of a besieging a city, were never short of imagination for ways to end it. To penetrate the ramparts of a city, it was sometimes easy to bring the enemy out by simply throwing over an effective weapon, such as a decomposing dead body. If the body is of a man who had died from an infection, and carries the disease-causing germs, it adds to the stench. Often it was even simpler to throw excrement, the effect of which on the other side would be equally devastating.

Thun-l'Eveque

The Hundred Years War began in 1337 when King Edward III of England staked a claim to the crown of France. Both, the English and the French independently negotiated with the Earls and Dukes. In 1338, Edward III sent a military expedition to the Continent. Flanders was then allied with the French and Hainaut with the English. In 1339, during the initial exchanges, the castle of Thun Levesque (today, Thun-l'Eveque), north-east of Cambrai was taken by Gautier of Masny, Duke of Hainaut, on behalf of the English. At the time this castle was located in Flanders, on the banks of the Escaut river bordering Hainaut. It was from here that Gautier and his garrison raided Flanders, particularly around Cambrai, which then turned to John, Duke of Normandy for protection. In 1340, the latter in turn, captured the castle. From Cambrai and Douai, he brought catapults to pound the walls and then besieged the castle with stones as well as bodies of dead horses and other animals. The defenders could not stand the stench and were ready to accept a truce or to surrender, if they did not receive help from the Duke of Hainaut within 15 days. Their demand was heard by the Duke, who put up a large army on the other side of the Escaut, to face the Duke of Normandy. However, the troops of the Duke of Hainaut, found it too risky to go across and finally ordered the garrison to abandon the castle.

Siege of a castle from the Middle Ages.

Siege of Karlstein

Karlstein (Carolstein) was founded in 1348 by Charles IV, King of Bohemia, Emperor of the Holy Roman Empire. The crown jewels and the precious relics of the saints were kept in a fortress. Towards the end of the 14th century, a huge reformist movement broke out, led by Jan Hus. Around 1386, at the University of Prague, Jan Hus studied the works of the English theologian, John Wycliffe which led him to radically question the establishment of the Church. He denounced its abusive and excessive wealth. His growing influence began to worry the authorities. Excommunicated in 1411, he went to the Council of Constance with protection from Sigismund, who was the King of Hungary and Emperor of the Holy Roman Empire, but who ultimately betrayed him. In 1415, the Council condemned Hus to death. Crowned with a paper miter with devils painted on it, he was burnt at stake, which scandalized his followers. The popular discontentment spread and soon turned into an explosive revolution among the poor masses of Prague, led by the priest, Jan Zelivsky and later by Procopius the Bald. In 1419, Wenceslas IV of Bohemia tried to reduce the influence of the Hussite movement by limiting it to a few churches. Wenceslas died a few weeks after taking this unpopular measure and his brother, Sigismund "the traitor" succeeded him. The Czech Nobles opposed his nomination because he denied them their religious freedom. In 1421, they offered the crown to Alexander Witold, Grand Duke of Lithuania. Witold was in a dilemma because if he accepted religious tolerance for the Czechs, he would be in conflict

Prince Sigismund Koribut.

From the Dark Ages to Modern Times

with the Pope and the Emperor (who had already organized two crusades against the heresy Bohemians, although unsuccessfully). He tried to stall by sending a regent of Bohemia, the Prince Sigismund Koribut, nephew of the King of Poland.

The Prince Koribut, a very good soldier and diplomat, gave his support to the reforms. He soon unified the Nobles and important Hussites and became regent of the Diet in 1422. Soon afterwards, he conducted the siege of Karlstein, the stronghold of Bohemia and the most important Catholic city in the neighbourhood Prague. In 1422, the castle was assailed for five months by the Hussite artillery. Unable to take Karlstein by force, Koribut catapulted bodies of soldiers killed during the siege as well as 2000 wheelbarrow-loads of dung, over the walls. This was the first use of products that were less bulky but virulent. According to the chronicler Anthony Varillas, the stench was so strong that "it knocked off the teeth of the defenders"! They would not have been saved but for the skills of a rich Bohemian pharmacist, who provided them medications, in return for a large sum of money. An armistice was concluded with Koribut releasing his troops for another threat, the Third Bohemian Crusade. Finally, the revolutionary Hussite movement was defeated in the battle of Lipany (1434) by an army comprising of the Czech nobility and gentry.

4 Plague

The word pest, comes from the Latin, *pestis*, meaning plague, and historically from the Sanskrit word *pyi*, meaning epidemic or harm, which is entirely appropriate. Indeed, in the history of mankind, no other calamity has had as devastating an effect and been charged with such intense drama as the plague. It has also made a lasting impression on our collective conscience. In recorded history, an estimated 200 million individuals have been victims of the plague. It is caused by a tiny bacterium called *Yersinia pestis*, which is transmitted by the bite of fleas infesting common animals like rats.

Mithridates of Pontus, "the poisoner"

Mithridates VI Eupator (called the Great) (around 132- 63 BC), King of Pontus, located on the southern coast of the Black Sea, began a political conquest in 111 BC. An oriental Sultan of a Hellenist culture, he wanted above all the prosperity of his Kingdom and his policy was to chase Rome out of Asia. In 74 BC, Mithridates began a long siege of the city of Cyzicus on the Back Sea. But the inhabitants defended it energetically. As the siege continued, the troops began to feel the effects of hunger and disease. Corpses now appeared in the streets and this coincided with the arrival of the plague. Mithridates was forced to lift the siege. We do not know if the plague broke out "naturally" among the troops of Mithridates or

Mithridates (Louvre Museum, Paris).

if the defenders had deliberately thrown plague-infected corpses to spread the disease, as did the Mongols of Golden Horde later on, during the conquest of Caffa.

Mithridates was finally defeated by Pompey in 66 BC. Interestingly, retreating in Crimea, he made several attempts at suicide with his own poisons, but failed as he had become immune or "mithridated" to them. Finally he ordered Bituitus, one his Gaul soldiers, to kill him, as he did not want the enemy to use against him the weapon that he had himself used extensively: the poison!

The first pandemic or the Plague of Justinian

While the Roman Empire was disappearing from the West in 476, it put up a resistance in the East, primarily in the city of Constantinople named after its Emperor Constantine. In the 6th century, this city had taken its name of Byzantium, in the memory of the Greek village that existed earlier. After the occupation of the Empire by the Turks in 1453, the sultans made it their capital under the name of Istanbul.

In 518, after the death of the Emperor Anastasius, the senate chose as the successor, an old, illiterate soldier called Justin, who had risen to become Chief of Security. Pretty soon he adopted his nephew, Petrus Sabbatius, hailing from what is today the city of Sofia, who took the name of Justicianus or Justinian. After the death of his Uncle, on the 1st August 527, Justinian, at the age of 46, took the reins of the Empire to rule brilliantly until 565. After having defeated or suppressed the Barbarians and the Persians, he proceeded to regain parts of the Mediterranean, the old Western Empire, that is to say, the Italian peninsula, Spain and the maritime façade of Northern Africa. Once a firm control over this region was assured, a new civilization, called the Byzantine, named after the capital, took birth during this period of prosperity. However, like a harbinger of divine fury, there was a series of disasters: an outbreak of plague, an earthquake that destroyed most of Antioch, burying alive 25,000 inhabitants, a comet that obscured the sky for a whole year (!) and a famine in Italy. Following these ominous signs, was the first authentic epidemic recorded in history, the Plague of Justinian.

The infection seems to have originated in Ethiopia, in the region of Lake Victoria which had been associated with epidemics of plague since long. It is likely that the waters of the Nile, flooded after heavy rains, carried the remains of the natives' huts including the rats inhabiting the reeds (used in making the roofs), all the way to Egypt, thus causing a severe outbreak of plague. Alternatively, the plague could also have been imported from Central Africa, that is to say, Uganda, Kenya or Congo, in the caravans that brought slaves and ivory.

Saint John Chrysostom, called Golden Mouth, a contemporary of this epidemic, reports on the Christian stylites ascetics.

Emperor Justinian
(San Vitale Church, Ravenne).

These hermits lived in 10 meter high columns, built for prayer and meditation. They were in contact with the outside world only through a basket that would be lowered by a rope, for their disciples to place their offerings. None of these "saints" were affected during the epidemic, although there would be innumerable corpses lying around their columns. Indeed, this observation had the effect of inducing several people into this vocation, leading to a significant increase in the number of stylites in the 6th century. One can easily explain this "miracle" today: in order to infect the hermits, the fleas that transmits the causative bacterium, *Yersinia pestis*, needed to survive in the baskets, but in fact, they cannot live for long without body heat. The stylites were thus spared.

Pelusium is an ancient city in Egypt, located on the Eastern mouth of the Nile, surrounded by lagoons and swamps. From here, the plague spread rapidly through Syria and Asia Minor. Byzantium, was an important city that relied on wheat imports coming from the Mediterranean basin. Ships that transported wheat, also carried the rats that brought the disease to Byzantium, which soon became depopulated in 542. It is estimated that in 543 alone, half the population numbering about half a million was decimated. Going up the Danube, the epidemic gained ground in Hungary, Austria and Italy and finally up to the regions of Reims in France, and Trier in Germany. Eleven outbreaks occurred between 558 and 767, following a cycle of about 8-12 years.

Byzantium experienced a resurgence of the disease in 580 and Justinian himself almost succumbed to this evil. The numbers sometimes appear exaggerated by the chroniclers, who spoke of 100 million dead, and some of the mortality rates correspond to 130% of the population! Nevertheless, this epidemic was severe as it is estimated that, depending on the region, 50-60% of the people were affected. However, all localities were not equally affected. The plague affected mainly the areas around ports and diffused more thinly in land as the connections between these regions were not so well developed in the early Middle Ages as during the second pandemic. It is therefore not surprising that it also left an impression in the minds of the people, as chronicled by Warnefried, that the plague "depopulated towns, turned the countryside into desert and made the habitations of men to become the haunts of wild beasts". Gregory of Tours (538-594) in his chronicle of the High Middle Ages Merovingian "History of the French" has written: "As there was shortage of coffins and planks, at times ten or more would be buried in the same pit. One Sunday, in the basilica of St. Peter's in Clermont alone, there were 300 dead bodies".

The reign of Justinian is synonymous with general prosperity. Nevertheless, it was not so in all the provinces, like the Balkans that had been neglected. From 540 this region began to attract the Slavic invaders who crowded under the walls of Byzantium and devastated Greece. Moreover, the issue of religion played an essential role in the weakening of the Holy Empire in the East, towards the end of the Dark Ages. Seventy years after the death of Justinian, in 565, countries like Egypt or Syria could not resist the invasion by the Arabs who were able to settle in the large areas that had been impoverished, depopulated and thrown into disorder by the plague. However, quite certainly, plague was not the only epidemic, as smallpox, another scourge of those times, also played an important role in the context of combined epidemics. For instance, King Dagobert I, escaped the plague but later died of the pox in 638.

From the 8th to the 14th century, Europe appears to have been spared from major epidemics, as supported by the fact that its population grew by 300% in this period of time.

The second pandemic or the Black Death of the Middle Ages

The second pandemic of plague most certainly originated in Central Asia. Caffa, located on the eastern coast of Crimea (now in Feodossia or Fedosiya) was an important trading junction. At the intersection of the Silk Route that was the major caravan route through China, and the Spice Route that connected with India, it was a thriving economic center. Crimea was inhabited by the Tartars from the dry and treeless steppes. Jealous of the rich Italians, they took advantage of the rift between a Genoese (Christian) and a Tartar (Muslim) and established their headquarters in this city for three years starting in 1344. Plague hit the Tartar army which lost several thousands of its men. In 1347, the bodies of hundreds of soldiers who died of plague were thrown over the walls, by their chief, Khan Djanisberg, "so that the Christians would be annihilated by the stench". It should be noted, by the way, that this required machines more powerful than the catapults that were available at that time. The new machines were called trebuchet catapults, from the French word meaning "bird-traps" and ironically, its inventor was the first victim, because he was accidently thrown inside the citadel while making final adjustments to the eyepiece. Gabriel de Mussis, who reported this first act of "biological warfare" believes that these catapult launchings initiated the disease among the besieged. This hypothesis, attractive as it sounds, may not be quite right. In fact the hygienic conditions inside the village were already so poor that it seems more reasonable to think that it was a natural cycle that brought in the field rats and their fleas, the originators of "intramural" plague. In any case, the loss of human lives was so heavy on both sides that the siege was lifted. The Genoese merchants hastily loaded their boats and set sail in the direction of Italy.

It was in October 1347 that 12 galleys approached the port of Messina in Sicily. After two weeks of travel, half of the passengers had perished. Frightened by the tales of the survivors, the inhabitants of Messina replenished the boats and joined the occupants to return to the seas at the earliest.

The Black Death (Illustration in German Bible, 1411).

From the Dark Ages to Modern Times

Triumph of Death, by Brueghel the Elder (Royal Museum of Fine Arts, Brussels).

However, a few hours later, the first cases of plague were reported in Messina. All of Sicily was hit by the scourge while in the city of Catania there were no survivors. The disease appeared in Sardinia, and the city of Marseilles by All Saints Day in 1347, where the entire population was affected. In June 1348, the plague reached the valley of Rhone and at the same time spread through Italy, South of France and the East of Spain. Paris was hit and lost 80,000 inhabitants in a few months. In 1349, England and Germany were contaminated in their turn. In 1350, it was all of Northern Europe that was affected; Iceland lost half of its population. In five years, an estimated 17 to 28 million deaths occurred in Europe, which had a population of 50 – 60 millions. A third of the World population was decimated by the plague. Several generations were required to rectify the disastrous effects of this horrible "Black Death", so called due to the dark hemorrhagic patches that covered the bodies of the diseased. The spread of the epidemic stopped at the boundaries of the very hot or very cold regions, which was inexplicable at that time. What the contemporaries attributed to "divine spirits" may now be attributed to three successive, abnormally cold and wet summers. The year 1348 saw an unrelenting rainfall from 15th August up to Christmas. Under such weather conditions, the crop production was very low, leading to famine and hence a greater susceptibility to infections. After studying the plague in Avignon, Guy de Chauliac has described well its different forms: "the said ailment was of two types: the first lasted 2 months, consisting of continuous fever and spitting of blood, and from this, one died in 3 days. The second lasted for the rest of the period, also with continuous fever and apostumes (abscesses) and carbuncles on the external parts and from this, one died in five days. The contagion is so great (especially the one with the coughing of blood) that it spreads not only by contact but simply by looking at each other... A father no longer visits his son, nor a son his father; charity is dead and hope is defeated".

Facing a grave threat such as this, human behavior often reverts to the primitive instincts. Some made great sacrifices (out of the 24 doctors in Venice, only 4 survived, which shows their dedication to the cause) while others not having the same attitude, even took refuge in debauchery. In his "Decameron" Boccacio (1313 – 1375) writes about the epidemic that gripped Europe around 1351 – 1352: "From the fine city of Florence, the beauty of which exceeds that of any other Italian city, came a deadly stench...The disease continued to spread, getting worse, because not only speaking with or visiting the sick that put the healthy ones at the risk of falling sick or even of death, but it was enough to simply touch clothes or other objects that had been touched by the sick for this mortal disease to be passed on. This made them abandon the sick and instilled such a terror in the hearts of men and women, that a brother would leave his brother, an uncle his nephew and often a wife her husband...Even parents were afraid to tend to their children." Or "Corpses were stored in the mass graves like bales in the hold of a ship. Everywhere, cemeteries had to be expanded urgently, or multiplied". In the South of France, corpses were thrown in the Rhone, the Pope of Avignon having blessed them so that they received a Christian burial. All strata of the society were affected. Even though the rich could escape to their residences further away, they would nevertheless carry with them the fleas infested with the plague bacteria in their hair and clothes, their hygiene being so poor.

This terrible disease served as a reason for persecution of the Jews. In fact, because of certain peculiar habits dictated by Talmud, the Jews followed a personal and community hygiene that was quite developed. Also, compared to the other parts of the town, their neighborhoods had

Bonfire of Jews in Cologne during the plague (Wood carving, Liber Chronicarum Munbis, Nuremberg, 1493).

much fewer rats, the carriers of the disease. Moreover, the medications that they gave their patients were plant-based and did not involve purgatives or excessive bleedings performed with contaminated instruments. As a result, the Jewish population was comparatively protected by the plague.

This fact was not much appreciated by the clergy who used this pretext to accuse the Jews of being the cause of the disease. Had they not seen them contaminate rivers and wells by throwing mysterious poisons made up of droppings of spiders, snakes and

From the Dark Ages to Modern Times

owls? It was in 1348, in Chillon on the Geneva Lake that the persecution began. In Freiburg, all the Jews were confined inside a wooden building, which was set on fire to burn them alive. In Strasbourg they were hung in the Jewish cemetery in the city. As this movement spread, a large number of Jews moved to the Eastern parts of Germany or to Poland, northwestern Austria and western Russia, settling in large communities.

Despite the persecution of the Jews, the plague continued to spread, making the church authorities realize that they were not in fact responsible for the scourge. In 1349, Pope Clement VI issued a bulletin condemning any acts of violence against the Jews, even welcoming them into the city of Avignon.

The "Flagellants" also appeared during this plague. This Brotherhood founded in 1260 in Perugia by the Dominican monk, Ranieri Fasani, comprised of men stripped to their waist, who lashed themselves with a belt for 33 consecutive days to atone for the sins originating from the divine wrath. Men who were true to their initial Brotherhood were soon joined by murderers, thieves and all those trying to evade justice. By the end of 1349, it is estimated that they numbered nearly one million in Northern Europe. Papal authorities and princes, alarmed by their severity decimated them mercilessly.

Flagellants (Wood carving, 15th century).

It was also at this time that Saint Roch, a native of Montpellier left for Northern Italy to care for the sick. He was himself infected, but survived miraculously. However, on his return to Montpellier, he was suspected of espionage and confined to a dungeon.

The plague went up all the way to Greenland, affecting the colonies founded by Erik the Red in 936. The weakened men, not being able to return the onslaughts of the Eskimos, disappeared, marking the end of the Viking colonization. Their connection with the "Vinland" situated on the coast of Canada and Newfoundland was thus destroyed, and this most certainly changed the course of pre-Columbian American History. Notably, after this time, Greenland was altogether forgotten until its rediscovery by John Davis in 1585.

The Black Death thus caused an abrupt halt to the steady population growth that had begun 5000 years BC. It took 150 years for Europe to regain its initial population.

The plague then began its long drawn out decline up to the 18th century. Nevertheless, there were a few dramatic upsurges as in 1437 in Paris (50,000 dead), or in 1466 in Constantinople (600 dead per day), or in 1628 in Lyon, France. On the 20th June 1720, in a house on the Rue Belle-Table in Marseilles, a washerwoman, Marie Dauplan, aged 58 years, died after

a brief agonizing illness: she was the first of the 100,000 victims of the fresh incidence of the epidemic. In 1799, the plague made one of its last appearances when it attacked Napoleon's army, killing 4,000 of its ranks while around 3,600 were killed in combat.

The plague in Marseilles in 1720 (detail) by Michel Serre, Fine Arts Museum, Marseilles.

On the 28 April 1828, Russia declared war against Turkey. The conflict took place in the Balkans, the Walachia, Bulgaria and parts of Transcaucasia. In the west, the Russians captured Varna and retired to the left bank of the Danube. During the second campaign in 1829, Diebitsch defeated the Turks at Kulevtchi and then invaded the Balkans up to Adrianople where the Turks surrendered without resistance. It was during this campaign that an epidemic of plague that had spread through Asia Minor in the year 1828, reached the western parts of Turkey and Walachia. It had appeared sporadically in Bucharest in 1825, infecting the Russian troops there. After May 1828, the plague propagated through Walachia and then all over the country by autumn. During this campaign, a total of 210,000 Russian soldiers died of various infections including, malaria, diarrhea, dysentery, various fevers and the plague.

Ivan Dibich (Hans Karl von Diebitsch) by George Dawe.

The Black Death had considerable economic, social and moral consequences on the medieval society. The depopulation of the countryside led to a profound and long lasting impact on the development of the western society. A morbid fascination for Death ensued and there was a proliferation of paintings called "macabre dance" depicting individuals from all sections of the society, religious or not, being courted by death.

The disease disappeared from Europe gradually, with the last cases in Turkey in 1841. But one question still remains unanswered: how did the epidemic subside? Did it have to do with a climatic change, cessation of the heavy rains of 1348, better hygienic measures, use of soaps that removed the fleas, the newly introduced habit of undressing before sleeping, which reduced the proliferation of the fleas and lice which require a temperature close to the body temperature to multiply. The hypothesis of a better hygiene explains the simultaneous recession of the plague and typhus.

There is another popular hypotheses, based on changes in the murine population. Indeed, the species of rat, *Rattus rattus* (house rat or granary

From the Dark Ages to Modern Times

rat or black rat), the main carrier of the disease, was replaced by *Rattus norvegicus* (gray rat) that is more resistant to the plague and is less in proximity with humans. Also, the two species are fiercely antagonistic. However, others refute this hypothesis because the gray rat did not invade Europe until the first half of the 18th century, that is to say after the decline in the epidemics had already begun. There was an earthquake in Asia that caused a passage of hordes of gray rats to Europe and later to the rest of the world, after crossing the Volga. However, the European black rat has not completely disappeared.

Rats and the plague.

Another hypothesis links the decline in the disease with the massive use of arsenic since 1650, in an attempt to control the rat population.

References to the plague can be found in several historical and literary works, such as the two given below.

Romeo and Juliet, by Ford Madox Brow, (Museums and Art Gallery, Birmingham).

Romeo and Juliet, victims of the plague

After their secret wedding, Romeo was separated from Juliet and was in exile in Mantua. In order to prevent her family, the Capulets, from marrying Juliet by force, Friar Lawrence made her drink a potion that made her appear dead although she was only unconscious. Unfortunately, the letter written by Friar Lawrence explaining the plot to Romeo, never reached him. As luck would have it, Friar John who was entrusted the letter, was forced into quarantine due to the plague that was raging in Mantua. Shakespeare (Act V, scene 2) describes it as "where the infectious pestilence did reign".

Siege of Reval

From the beginning of the 17th century, the kings of Stockholm wanted to transform the Baltic into a "Swedish Lake". In 1660, they succeeded in signing a peace treaty with Prussia and Poland. This only exacerbated the mutual hate and envy. In 1697, a young King Charles XII, just 15 years old, ascended the throne of Sweden. In three years, he defeated the Danes and put the Russian army in disarray by killing 15,000 soldiers and taking 20,000 prisoners in Narva in the Gulf of Finland. Having learnt a lesson from this bitter defeat, Peter the Great, the Russian Tsar completely revamped his army and modeled it in the Western style.

Charles XII of Sweden, by David von Krafft (National Museum, Stockholm).

For several years, Charles XII, "Alexander the Great of the North" knew only victories on the battlefield. However, after 1708 – 1709 the success gave way to defeat. On 27th June 1709, the armies of Charles XII were crushed at Poltava by a 42,000 strong, modern Russian army. Taking advantage of this, the armies of Peter the Great seized several ports in the Baltic, including Reval (today, Tallinn), the major port of Estonia, located on the coast of Gulf of Finland. During the siege of Reval, the Russians used plague-infected corpses to contaminate the city held by the Swedes. Finally, Charles XII died in December 1718 during the siege of Frederickshald in Norway under rather obscure circumstances.

The third pandemic of plague

This initiated in 1855 in the west of Yunnan in China and then diffused up to Yunanfu (Kunming), Canton (1866), then by Hong Kong (1894). This pandemic propagated faster than the previous ones because of traffic of trains and steam boats. It hit Mumbai (earlier, Bombay) in 1896, Suez in 1897, Alexandria and Portugal in 1899, and then in Marseilles, Glasgow, and Hamburg, where it ended abruptly. Other continents that had been spared so far, such as America and Australia, were also affected. The plague caused several hundred deaths in the Chinatown in San Francisco, where it was introduced by a merchant ship coming from China carrying with it infected rats (one clandestine passenger died from plague during the journey). Later, in 1910 – 11 the plague appeared in Manchuria; and in 1920 – 21, in Shansi, where two epidemics took 60,000 and 16,000 dead. In Doukkala, Morocco, it took 15 000 victims. This pandemic was particularly deadly as it killed one million people in India in 1903 alone, while a total of 12.5 million Indians fell victims during the period 1898 – 1918.

It was during this third epidemic that Alexander Yersin of Institute Pasteur, France, on a mission in Hong Kong, isolated the causative agent, a bacterium that was to be named after him, *Yersinia pestis*.

Bringing back the body of Charles XII, by Gustav Cederstrom (National Museum, Stockholm).

From the Dark Ages to Modern Times

5 Other important epidemics and their impact on History

Malaria

Malaria is one of the major causes of mortality today, causing about 2 million victims each year, especially, children under 5 years. It is estimated that every ten seconds, a person dies of malaria in Africa. This disease most certainly originated in Africa around 3 – 5 thousand years BC and was passed from apes to man. The infection spread in tropical Africa up to the valley of Nile, in Mesopotamia and then the Mediterranean coast. Greek merchants seemingly helped its spread in Sicily and the south of Italy.

An ancient disease, malaria has been mentioned in the papyrus of Ebers (1570 BC), on the clay plates from the library of the King Ashurbanipal (692 – 627 BC) and in the classical Chinese literature of Nei Chang (2700 BC). Basically a local disease, it turned into an epidemic due to deforestation and the appearance of areas of stagnant water, which serve as breeding grounds for the disease vector, the Anopheles mosquito in which the parasite, *Plasmodium* multiplies. Under these geopolitical circumstances, malaria was able to reach the Mediterranean basin and spread rapidly.

Fall of the Roman Empire

The areas located around Rome were dotted with malaria-infested swamps in which the Anopheles mosquito proliferated. In the year 79 AD, soon after the eruption of Mount Vesuvius, there was a huge outbreak of malaria in Rome. The infection appears to have been

Europe in 476.

localized in Italy alone, causing numerous victims not only in the cities but also in the countryside, including parts of Campania, the "garden" of Rome, which remained infested for a very long time. During the first few centuries of the present era, many Roman farmers were forced to seek refuge in Rome leaving behind their fields uncultivated and neglected. This resulted in an even greater dependence of the Empire on their colonies. Moreover, since the Italian population was severely affected, Rome was forced to reconstitute its legions with mercenaries coming from the Germanic tribes that it had conquered. The Roman legions thereby lost their original spirit of cohesion. Population movements contributed to the destabilization of the Roman Empire that experienced its downfall towards the end of 476.

Dissemination in the New World

The official medallion of the British group against slavery (Josiah Wedgwood, 1795).

Between the 15th and 19th centuries, an estimated 12 – 20 million Africans were subjected to slave trade in the New World. The regions involved were primarily from West Africa where malaria and yellow fever were rampant. The local population from these regions having been exposed to these diseases for centuries were relatively protected from them. In contrast, the white Europeans and especially the native Americans succumbed to them.

Yellow Fever

Yellow fever is caused by a virus transmitted by the mosquito of the genus *Aedes*.

Haiti

The first dramatic incidences of the disease occurred in the Caribbean, particularly in Haiti. The Black population, originally from Africa were able to tolerate the infection better, but the White population was highly sensitive and suffered immensely. In the 18th century, the island of Saint-Domingue, today Haiti, was still under the French authority. In 1794, during the French Revolution, English troops embarked and seized Port au Prince. The English encountered a very tough resistance from the black population led by Toussaint Louverture. In addition, the yellow fever which was endemic in this country at that time took 7,500 victims from the British troops, forcing their retreat. In order to win the island back, Napoleon then sent an army of 20,000 soldiers led by general Charles Victor Leclerc, who was his brother-in-law and the husband of Pauline Bonaparte. Although the insurgents were forced to surrender and their leader, Toussaint Louverture, was captured, a new outbreak of yellow fever annihilated 23,000 French soldiers and General Leclerc himself in November 1802. Completely dejected, the French pulled out from the island and on 1st January 1804 the first independent republic of Latin America was created.

Toussaint Louverture (French carving, 1802).

European acquisitions in North America in 1800.

The colony comprising of 520,000 inhabitants in 1789 was the most populated and the richest in the French Caribbean. In 1804, after the departure of Toussaint Louverture in captivity, the new chief of the insurgents, Jean-Jacques Dessalines (who called himself, Emperor Jacques I) ordered a general massacre of all the Whites who were still living on the island, such that none escaped. The population thus reduced to 10,000 mulattos and 230,000 Blacks.

As a result, many of the French established themselves in New Orleans, USA. The repeated exchanges between their adopted city and Saint-Domingue was the basis of numerous epidemics of yellow fever between 1790 – 1805 in New Orleans. After Saint-Domingue became independent under the name of Haiti and after it got rid of Europeans, the contacts with this island became less frequent and eventually the yellow fever epidemics in New Orleans declined.

There were other important consequences of the departure of the French from the island. Soon afterwards, the US President, Thomas Jefferson sent emissaries to Paris with the Minister of War, Talleyrand, to investigate the possibility of buying a few cities in Florida which were controlled by the French. To their surprise, they learnt that Napoleon

Talleyrand, by Francois Gerard
(Chateau de Valencay).

had completely given up hope of settling in this part of the World, as he required his troops for future conquests in Europe. Hence, for a mere pittance, they could acquire an area equivalent to a third of the present United States of America. Indeed, as Louisiana represented all of the central part of USA, its acquisition allowed the settlers to make their migration westward during the so called "conquest of the West".

Panama Canal project

Similarly, in 1889, the Panama Canal project, developed by the French under the authority of Ferdinand de Lesseps was completely abandoned due to the high mortality caused by yellow fever that prevailed among the workers. It was only in the 20th century, through the efforts of Major Walter Reed of the American army that the infection could be controlled.

Syphilis

Syphilis is a sexually transmitted disease caused by the bacterium, *Treponema pallidum*.

The first reports of the disease

On the 3rd August 1492, with orders from Christopher Columbus, 120 men embarked three caravels and sailed from the port of Palos, near

The convention of Mortefontaine. Napoleon signing the acquisition of Louisiana by the United States.

Seville. After tackling Cuba and later Hispaniola (Haiti), they returned to Seville on the 31st March 1493 (in fact only two of the three caravels returned, as one was wrecked!). A second expedition was planned and set sail on the 2nd February 1494. The first incidences of syphilis in Spain coincided with these expeditions. The correlation between the sexual transmissibility of the disease and the fact that the sailors of Columbus had sexual relations with the Caribbean women was soon established. In 1539, Ruy Diaz of Seville wrote in his "Serpentine Malady" that "as the crew of Admiral Cristobal Colon were involved in trade and relations with the Indian women, the disease was easily transmitted to the entire crew and the symptoms were obvious in the Squadron - on a pilot from Palos,

From the Dark Ages to Modern Times

called Alonzo Pinzon and on the others who suffered from it". Pinzon was the first recognized patient of syphilis in history. He died soon after his return, his body covered with buboes. In fact, it is hard to say if it was a really new disease imported from across the Atlantic, but on noting the suddenness and the speed with which it spread, in the late 15th century, one cannot deny the correlation between its appearance and the journeys of these sailor. Moreover, a recent genetic study has indeed demonstrated the origin of the disease to be in America.

The admiral Colombus, by Ridolfo Ghirlandaio (Sea and navigation Museum og Gena.

Bartholome de las Casas, an eye witness to the arrival of Christopher Columbus in Seville recounts: "it is also a proven fact that the incontinent Spaniards did not observe the virtue of chastity and hence had the buboes; and that in a hundred, not one was spared, except in the cases where their partner ("la otra parte") did not have the buboes. The Indians, men and women, who were affected, suffered little, or a little more if they had the pox (virulae), whereas the Spaniards were in great pain and their torment continued relentlessly during the period of the buboes".

Arrival of Christopher Columbus in America (Library of Congress, Washington).

Conquest of Naples

While the new disease was breaking out in Spain and Italy, Charles VIII, the King of France, decided to conquer the kingdom of Naples (southern Italy) with the help of his army comprising of 30,000 mercenaries coming from France, Germany, Flanders, Poland, England as well as other countries. The troops entered Naples on the 12th February 1495. Fallopius, the celebrated Italian anatomist, whose father was in the citadel reports, "they (the besieging troops of Naples) drove out of the citadel all the whores and women, especially the most beautiful ones, who they knew had been afflicted by some contagious disease, under the pretext that provisions were limiting. The French taking pity on them and being allured by their beauty welcomed them". A Holy League was formed by the Italians to expel the French, but in fact Charles VIII was forced to return to the North because his troops were dying of syphilis. Naples was lost.

In 1495, the troops were remobilized to Lyon, where the Hotel Dieu became the first hospital to specialize in the treatment of syphilis. Incidentally, one also notes that this was also one of the first conflicts during which the belligerents made a deliberate attempt at bacteriological warfare. In fact, the Italians would purposely leave bottles of wine that had been contaminated with blood from leprosy patients so that the French would contract the disease. This was a totally wasted effort, because leprosy does not spread by the oral route!

Arrival of Charles VIII in Naples 12th February 1495, by Eloi Firmin Feron (second quarter of the 19th century. National Museum at the Chateau de Versailles and de Trianons).

The sickness of Naples, Albrecht Durer (1496).

Syphilis is known by many names. As they say, it is always the neighbor's fault! Thus, the disease has been variously called "mal de Naples" by the French; "mal francese" by the Italians; "mala de frantzos" by the Germans; "French pox" by the English; "Spaanse Pocken" by the Flemish and the Dutch.

Europe was rapidly contaminated by syphilis as the soldiers returning from Naples dispersed. Only the northern regions such as Finland and Lapland were spared for a while. Similarly, the Albanian and Roman mercenaries serving in the Venetian army, and the Swiss and German foot soldiers, all contributed significantly to the spread of the disease when they returned home. French chroniclers have written about the Burgundians (500 horsemen and 700 infantry), most of whom were suffering from the "mal de Naples" were not allowed to enter the city Metz in May 1495. However, the soldiers managed to infect some whores who later introduced the disease in Metz, where it lasted for four years. From Europe, the disease then spread to Africa, then to Middle East and Asia, as a result of people movements and developing trade relations.

It was not just the sailors and soldiers that were affected by syphilis but some of the promiscuous royalty and celebrities as well. Their performance and decision making abilities were likely influenced due the effects of the infection, as we see in the following examples.

Alexander Borgia

Pope, Alexander Borgia (1431 – 1503), originally from Spain, is known in history as the author of the Bulle issued in 1493, which divided the New

World between Spain and Portugal and for ordering the execution of the Florentine reformist, Girolamo Savonarola, in 1498. He is remembered as man given to worldliness and corruption. He was also the father of the famous Lucrezia Borgia.

When Alexander Borgia displayed the first signs of syphilis, the physicians tried on him a form of therapy that was quite popular at the time. This consisted of plunging the patient, stark naked, into the belly of a bull that was still alive. Needless to say, this therapy was not really effective in curing syphilis!

Pope Alexander VI Borgia. Detail of the Fresco of the Resurrection, by Pinturicchio (1492 – 1495).

Francis I

The King of France who "had taken a mistress at the age of ten" was quite easily one of the first monarchs to be affected by syphilis. Madame Ferron, called "la Belle Ferronniere" or "the lovely ironmonger", wife of Jean Ferron, an advocate in the Parliament in Paris, was one the King's mistresses. Legend has it that the jealous husband frequented whores in the Capital in order to contract the disease with which he could infect his wife and in turn her lover, King Francis. However, this is nothing but a folklore, because Francis I had already contracted syphilis well before this episode. He had even passed on the infection to his wife, Queen Claude of France, who died of it in 1524, as well as to Anne de Pisseleu, duchess d'Etampes. Ironically, it was the King who infected "la Belle Ferronniere", and who in turn gave her errant husband the royal gift!

Francis I, King of France, by Jean Clouet (Louvre Museum, Paris).

Henry VIII

King Henry VIII of England, began his rule wisely in 1509, although he was just 19 years old. During the year 1527, he started to show signs of imbalance, quite certainly related to an infection of syphilis. Some authors have attributed the enlarged nose in his portrait by Hans Holbein to a lesion typical of tertiary syphilis. When Henry VIII fell in love with Anne Boleyn, he tried to discredit his legitimate wife, Catherine d'Aragon, who had not given him a viable male child. This affair had international consequences because the nephew of Catherine, the

Henry VIII of England, by Hans Holbein (National Gallery of Art, Rome).]

Germanic Emperor of the Holy Roman Empire, Charles V, pressurized the Pope to not annul the marriage of his aunt and the King of England. The Pope's procrastination incurred the wrath of the King, especially against his own advisor, the all powerful Cardinal Wolsey, who was until then in charge of State Affairs and whom the King now considered as an enemy of Anne Boleyn. Cardinal Wolsey was persecuted and died in 1529, after being accused of transmitting syphilis to Henry VIII by whispering in is ear! The rest of the reign of Henry VIII was a series of appalling acts until his death in 1547. For example, in 1531, he authorized a horrible punishment: death by scalding. From 1534, he was responsible for the massacres of Lollards, Lutherans and Catholics. This was followed by the beheading of Thomas More in 1535, and then the assassination of Anne Boleyn.

Ivan IV the Terrible

Ivan IV the Terrible, by Viktor Vasnetsov (Tretyakov Gallery).

In 1535, on the death of his father Vassili III, Ivan IV ascended the throne of Russia, although he was barely three years old. He was crowned Tsar on the 16th January 1547 and married a pious woman, Anastasia. In first years of his reign, he expanded his empire considerably. As he had quite certainly contracted syphilis before his marriage, two of his children died shortly after birth and the third, Feodor, was described as "stupid and retarded". After the death of Anastasia in July 1560, the Tsar began to display signs of dementia. All through his life he committed gruesome atrocities such as, those for example in the city of Novgorod, where he massacred thousands of people over five days, under the suspicion of conspiracy. Similarly, he ordered executions in Moscow in July 1570. Later, in November 1581, he himself slew his eldest son Ivan. On his death on 18th March 1584, the country was left in utter chaos. Shortly afterwards, Romanov took over the throne.

Assassination of the son Ivan by Ivan IV.

From the Dark Ages to Modern Times

Smallpox

Smallpox is a highly contagious disease, caused by a virus belonging to the group of Orthopox viruses and the family Poxviridae. This family which also includes the vaccinia virus, which is responsible for cowpox and the monkeypox and is potentially transmissible to man. Before the era of vaccination, smallpox was a threat to the entire human population. It is estimated that hundreds of millions of people have been victims since the oldest records of the disease. During the course of the 16th and 17th centuries, the disease has claimed several kings and queens, including the emperors of Japan and Burma, disrupting dynasties and the line of successions, thereby reshaping the States. In the 20th century alone, smallpox has taken 300 million victims, which is three times more than all the wars in that century.

Ramses V
(OMS, Geneva).

Origin of smallpox

It is believed that smallpox originated in Mesopotamia and spread through Egypt, where one finds its traces in few of the mummies, like those of Ramses V (died in 1157 BC). However, more sophisticated studies on the mummies using an electronic microscope did not reveal the presence of the viral particles. It is now believed that it originated in the Far East, 10,000 years ago and spread through China. The disease is described in the ancient medical texts from China and India, dating back to before 1100 BC. Some of the infections are sometimes attributed to smallpox such as those found among the Hittites in 1146 BC, or in Syracuse in 595 BC, or in Athens in 490 BC. The "epidemic of Antoninus" or the Antonine plague that came from Mesopotamia and hit Rome in the first century AD is generally attributed to smallpox (see the section on "Rome and the epidemics of plague" in Chapter 2).

Smallpox in ancient times

In 570 AD, a Christian army coming from Abyssinia (today Ethiopia) attacked Mecca, the capital of the Arabic peninsula with the aim of destroying the Kaaba and converting the local population to their faith. At this time, Kaaba was already recognized as a sacred place for the Arabs who were not yet Muslims. According to the Koran, God sent swarms of birds that threw stones at the armies of the enemies, which resulted in pustules that were contagious and spread like wild fire. The Abyssinian armies were decimated and their chief, Abraha himself died of the disease.

Along with malaria, smallpox is considered as one the major reasons for the fall of the Roman Empire. Marc-Aurelius himself was its victim. It appeared in Europe during the 7th century and was introduced by the

Sarascens. This disease was confused with others such as the plague and measles for many centuries. It was in the 7th and the 8th centuries that it raged as the deadliest epidemic, killing about 10% of the European populace. It was only after the experiment conducted by the English physician, Edward Jenner, on 14th May 1796, that the technique of vaccination was developed. In October 1977, a Somalian, Ali Maow Maalin was the last patient of smallpox, and he was cured.

Smallpox in the New World

Hernan Cortes.

The disease spread all over Europe where the people were more or less accustomed to its presence. It made a dramatic entry into Central America and in the Caribbean in the 16th century amidst people who had never been exposed to this virus. It is estimated that the population of Mexico went from 30 million to less than two million in the period of fifty years following the Spanish invasion, primarily due to smallpox and measles. North America had the same experience during the century following the arrival in Maryland, of a ship carrying infected people from England.

Panfilo de Narvaez left Cuba on the 23 April 1520 headed for Mexico to capture Hernan Cortes, whose disobedience and greatly displeased the Governor. He landed at Veracruz with 800 soldiers, 50 horses and artillery. They suffered losses in the first clashes with the natives before facing the troops of Cortes. Many of the soldiers from the Narvaez camp rallied to Cortes. Eventually, despite their numerical superiority, the Narvaez troops were defeated at Zempoala (Veracruz) on 24th May 1520 and Narvaez himself was wounded in the battle. He was taken to the Port of Veracruz and thrown in jail for 12 years.

The troops of Narvaez, including a certain number of African slaves, were shifted to the West Indies. By the time they landed on American soil, some of them had fallen sick and one showed clear signs of smallpox. The terrible infection then spread throughout the Aztec empire killing about half of the population of Mexico. It soon reached Central America, then South America and ravaged the Inca empire.

In 1532, about a dozen years after these events in Mexico, another Spanish conquistador, Francisco Pizarro arrived in the Inca empire. He captured the head, Atahualpa and kept him prisoner for 8 months. In return for the release of Atahualpa, Pizarro demanded a ransom of gold. However, on receiving the gold, he did not keep his promise, but executed his hostage. During these conflicts, an epidemic of small pox had largely contributed to the Spanish victories. The disease had arrived here in 1526 by land via Peru.

Francisco Pizarro.

By the time Pizarro landed on the shores of the country in 1531, his task was made simple because the epidemic had decimated a significant fraction of the population.

Anglo-French wars in North America

Sir Jeffrey Amherst.

It was perhaps the first real, premeditated, microbiological warfare that triggered a smallpox epidemic and decimated the native Indian tribes of Canada during the clashes between the British "redcoats" and the French soldiers of Louis XV, fighting to gain a "few acres of snow" (according to Voltaire!). The French had abandoned the mission by signing the Treaty of Paris in February 1763. However, unaware of these developments the native Indians fled from the valley of Ohio. United under the leadership of Pontiac, the chief of the Ottawa tribes, they managed to recover a few of forts from the British. Only three forts were retained by the British, including Fort Pitt (aka Fort Duquesne). This fort was occupied by Colonel Henry Bouquet, who was entrusted with the task of quelling the Pontiac rebellion by Lord Jeffrey Amherst, commander-in-chief of the British forces in North America. In a letter exchanged between the two, Lord Amherst's suggestion was to "contaminate the Indians with blankets infested with smallpox". The British conspired to reverse the unfavorable situation in which they found themselves. After a round of negotiations on 24 June 1763, Colonel Bouquet gifted the Indian emissaries two blankets and a handkerchief contaminated with small pox. This deadly disease being hitherto unknown in North America, the native Indians lacked any kind of immunity against it. Thus, an epidemic of smallpox was planted and spread rapidly causing considerable damage to the rank and file of the native Indians.

A little later, during the American War of Independence, the revolutionary government sent its troops to conquer Canada which was controlled by the British. After taking Montreal, the numerically superior Americans attacked Quebec but were defeated by an epidemic of smallpox and had to beat a chaotic retreat after burying their dead.

Louis XV of France and smallpox

The disease which allowed future Louis XV to ascend the throne was also the cause of his death. The end of the reign of Louis XIV was dark and difficult. War followed by the exceptionally harsh winter of 1709 led to a famine. The meager harvests coupled with excessive taxes threw France into a despair. The edifice of absolutism was bursting at the seams and the French hoped for a new reign. At just this time an epidemic of smallpox struck and eliminated one after the other, three heirs to the

Louis XV, King of France, by Quentin de la Tour (Louvre Museum, Paris).

throne. It was thus left to the great-grandson of Louis XIV, the Duke of Anjou, a small frail boy to become Louis XV.

The smallpox of Louis XV is famous. There are two versions about how he contracted the disease. Most of the authors believe that the King, who was known for his debauchery, was infected in the course of a sexual relationship that he had with a 12 year old girl who was a carrier of the disease. According to the official version, which is more flattering, he contracted the disease while attending the funeral of a young girl who had died of the disease. In any case, by 26th April 1774, Louis XV developed clear symptoms of the disease while at the Petit Trianon, a chateau located in the grounds of the grand Palace of Versailles near Paris. He died on the 10 May 1774 at 15.30 hours, amidst a general indifference and even joy in some parts of the Court. In view of the catastrophic financial situation and the considerable disrepute suffered by the monarchy, the King's entourage did not dare organize a public funeral, but quietly took his mortal remains to the Basilica of Saint Denis. Smallpox was so feared, that in order to avoid any risk of contamination, the body was enclosed in a double coffin: the outer one of oak and the inner one coated with lead and sealed with a compound made of vinegar and a liquor of camphor. Further, the medical personnel that served the late King were replaced so as not to contaminate his grandson, the new King, Louis XVI.

Smallpox during the 19th century

On 14th May 1796, the English physician, Edward Jenner injected a small amount of a sample of cowpox from an infected cow into an eight year old boy, James Phipps. This was the glorious moment of the introduction of an efficient vaccine against the terrible scourge of smallpox. Gradually vaccination became popular in Europe and the United States. In 1805, Napoleon had his entire army vaccinated. However this habit did not last over the decades that followed. Thus, during the Franco-Prussian war of 1870 – 71, the Prussian troops had been immunized, but not the French. An outbreak of smallpox that occurred affected 200,000 French soldiers, killing 24,000. Similarly, during the siege of Paris, 18,000 French became victims of the virus. There was no loss of lives among the Prussians, who captured Paris on 28th January 1871.

During the American Civil War too, Black soldiers fighting alongside the armed Northerners were more affected by smallpox than the Whites and the spread of the disease was facilitated by the fact that vaccination had been neglected.

Typhus

Typhus is caused by the bacterium, *Rickettsia prowazekii*. It is transmitted to man by lice, ticks or mites that stick to cloths and need our body heat to survive. Lack of proper hygiene favors the disease of typhus which is why it is encountered particularly during wars and natural disasters.

Depending on the situation and the geographical location the disease has been variously called, typhus exanthematic, malignant fever of the army, military fever, jail fever, camp typhus, ship typhus or war plague. Indeed, its history is closely related to military events. It is generally believed that during the Peloponnese War, the 65 year old, General Pericles died of typhus while trying to protect Athens in 429 BC. This disease is sometimes confused with the plague and hence in the 16th century it was called "pesticula" or little plague. In the battlefields, typhus was particularly deadly.

Arrival of the Catholic Kings and their children in Grenada. They were accompanied by the Cardinal Mendoza, who was already known as "the third King of Spain". 16th century, Philippe de Biguerny. (Sacristi de la Capilla Real, Grenada).

In 1490, typhus appeared among the Spanish soldiers who had fought with the Venetian army against the Turks. They brought the disease to Spain via Cyprus, contaminating in turn the troops of King of Aragon, Ferdinand II the Catholic. The latter troops lost 20,000 soldiers of which 3,000 had been killed by the enemy and 17,000 (!) by typhus during the capture of Grenada from the Sarascens in 1498.

The disease was also seen during the religious wars in the 15th and 16th centuries and during various wars that ravaged Europe in the 16th – 18th centuries, conditions that favored the spread of the disease.

Franco-Spanish wars

The Kings of France, Francis I and Henry II and those of Spain, Charles V and Philippe II, were all engaged in fierce battles for forty years. In 1490, there was an outbreak of typhus in Lorraine (France) during a particularly close conflict between Rene, Duke of Lorraine and the inhabitants of Metz. Later, while armistice was declared on the 18th June, the epidemic spread in Metz in August, forcing the residents to flee. The Nobles retired into their castles and other people dispersed all over the countryside. Hence, a quarantine was applied but was ineffective and the infection spread over the whole of Lorraine and northern Alsace.

The miseries of War, by Jacques Calot.

In 1528, an epidemic of typhus broke out in the north of Italy and travelled south just at a time when a battle for the control of Naples was raging between the French on one side and the German and Spanish troops on the other. The mortality reached an appalling figure of 30,000 French soldiers and 60,000 civilians. In fact the French were on the verge of victory. But for the outbreak of typhus that ravaged the French army, Francis I may well have seized Naples in this battle and secured a firm hold over this region. However, owing to the outbreak of typhus, it was Charles V who won and went on become the head of the Holy Roman Empire.

When Henry II succeeded his father Francis I to the throne, he resumed hostilities against Charles V and seized the city of Metz in 1552 and installed a strong garrison of 10,000 men. In November-December of that year, Charles V laid a siege around Metz with a 80,000 strong army including Germans, Spanish and Italian troops. The winter was particularly severe and the men camped in tents, without sufficient provisions, the convoys of replenishments having encountered the garrison of Henry II stationed at Verdun. Deplorable sanitary conditions prevailed in the camp and led to the simultaneous outbreaks of dysentery, scurvy and above all, typhus. The army of Charles V lost so many men to the diseases that the siege had to be lifted. It was typhus that had facilitated his victory in 1528, and it was typhus again that betrayed Charles V in 1552. Throughout the 16th century epidemics of typhus devastated Spain. Its population dwindled from 8,200,000 to under 6,000,000 within the span of half a century.

It was typhus again, that in 1566 prevented the army of Maximilian II of Austria from beating the Ottoman Emperor, Suleiman the Magnificent in Hungary. The disease was called "morbus hungaricus" and Hungary, "the graveyard of the Germans". In 1741, the Austrian armies could not defend Prague against the French as 30,000 of their soldiers died from typhus. At the end of World War I (WWI), the armies carried typhus to their respective countries, which explains why between 1917 – 1923, there were 20 – 30 million cases of which 3 million died, with a higher toll in Eastern Europe.

From the Dark Ages to Modern Times

Typhus in the Americas

Hernan Cortes landed in Veracruz in 1519 and in a few months, conquered the Aztec empire. Unfortunately, the victorious Spanish army was affected by several diseases such as smallpox, measles, leprosy and typhus. In 1520, smallpox brought by the crew of Narvaez (see above) killed half the local population, amounting to 3,500,000 people. In 1531, it was measles that exterminated a large number of the native Indians. In 1545, a new disease, most certainly typhus, took 80,000 Indians. In 1576, typhus struck again and took at least 2 million victims.

The Thirty year War (1618 -1648)

This was probably the one of the bloodiest wars in history apart from the one in 1914 – 1918. Besides the atrocities of war, military and civilian populations died also from a host of diseases such as dysentery, smallpox, scurvy and typhus. During this time, the population of Wurtemberg dwindled from 400,000 to 48,000. The great "battle" in Nuremberg in 1632 did not take place for the lack of combatants! There were just too many deaths on both sides from scurvy and typhus. Indeed, in terms of relative proportion, Germany experienced its highest depopulation in this period. This is how the Holy Roman Empire collapsed.

Scene from the Thirty year War. Wallenstein crossing a village, by Ernest Crofts (Leeds Arts Gallery).

Typhus: conqueror of Napoleon

During the years 1793 – 94, typhus was brought to Germany by the French. It appeared in Frankfurt-on-Maine in May 1793, then in the Black Forest, Swabia, Bavaria and Regensburg by December 1793.

During the conflict between the Republican army and the people of Vendee (western France), who were loyal to the monarch, there was a terrible outbreak of typhus during the siege of Nantes by the Royalists. Prisons and hospitals were filled to capacity. Several corpses lay abandoned in the streets of the city. In late September, the epidemic hit the prison of Sainte Claire where people were packed like sardines. According to Inspector General Leborgne, "this house lacked everything – air, water, food, medicines, even the means for burying the dead". Without enough beds, nor hay, the prisoners laid down on the humid floor and were fed moldy bread and water. At the prison of Bouffay, he writes, "the dead, the dying, the sick and the newly infected prisoners lay on the same pallet! The dungeons spewed putrid miasmas and the lights were dimmed by fumes from these stinking sewers". Also, "the disease was so severe that out of the 22 sentries that stood guard at the prison warehouse, 21 perished within a few days and the Members of the Board of Health who had the sad courage to visit the place, were almost taken victims".

The hospitals were so full that 3 – 4 persons shared the same bed. Of the 300 people requisitioned by the Revolutionary Committee for digging graves, most became infected and several died. A total of 10,000 perished in Nantes in this epidemic. Subsequently it spread all over Europe, in Bavaria first, then Wurtemberg and along the banks of the Rhine. During the battles of Austerlitz (1805) and Jena (1806) a large number of people died in hospitals from typhus. The disease took thousands, or even tens of thousands of lives.

Several deplorable epidemics of typhus occurred during the long occupation of Portugal and Spain by the armies of Napoleon between 1808 – 14. The French lost an estimated 300,000 to typhus and 100,000 in action. A particularly dreadful epidemic broke out at Zaragoza, Spain, in June – August 1808 during its siege by the French and again between December 1808 – February 1809. Among the 100,000 inhabitants of the city, 54,000 died of typhus which forced the city to surrender.

Episode of retreat of the Russians, 1835 (oil on canvas) by Joseph Boissard de Boisdenier (1813 – 66). Museum of Fine Arts, Rouen, France.

From Spain, the disease travelled to France with the prisoners of war. Dax was the first city to receive it and the disease followed the convoys, for example in 1809 it reached Bourges along with the prisoners. The latter were crammed into the barracks and hospitals. Of the 563 Spaniards, 103 died of typhus.

The campaign of 1812 by the Russians and their disastrous retreat from Moscow were particularly marked by this disease. During the Battle of Ostrowo, Poland, the disease broke out affecting nearly 80,000 soldiers. During their retreat in the shelters, the soldiers were subjected to hunger, fatigue and glacial temperatures (-30°C). Under these impoverished conditions, typhus

From the Dark Ages to Modern Times

spread easily and killed several thousands of them. Out of the 30,000 soldiers that sought refuge in Wilnia (Vilnius), in Lithuania, 25,000 died, along with 8,000 civilians. Among the 36,000 men that were besieged by the Russians at Danzig (Gdansk), 13,000 soldiers and 10,000 civilians succumbed to typhus. In total, about 500,000 men that constituted the Grand Army, barely 30,000 escaped the disaster. The disease was brought to France in 1814 by those who survived and there were so many patients in the hospitals in Paris that they had to be evacuated and sent by boats along the Seine all the way to Rouen, where the epidemic caused more deaths. The defeat of Waterloo in June 1815 put an end to these epidemics. Nevertheless, the disease still persisted in the prison and labour camps and also in impoverished neighbourhoods.

Dysentery

This disease spreads through contaminated food and water and is characterized by severe abdominal pain and diarrhea.

Dysentery and the Kings of France

Many kings of France have suffered and even succumbed to episodes of diarrhea, that were probably due to dysentery.

Louis VI (1081 – 1137), called "the Fat" (in French, "le Gros") was so obese and he almost could not move. He died on the 1st August 1137 after a series of attacks of diarrhea. Louis VIII the Lion returned from the Third Crusade and died of acute diarrhea on the 8th November 1226. This expedition had also resulted in the death of 22,000 soldiers caused by an epidemics of dysentery and malaria. Louis IX, called Saint-Louis, died at Tunis on the 25th August 1270 during the Eight Crusade. It is often believed that he died of the plague, but this hypothesis appears doubtful. The chroniclers of this time have described Saint Louis to be "frail, slender, tall and rather thin". In 1242, during a war against the English at Poitou, he contracted malaria. In 1243, the confessor of the his wife, Queen Margaret de Provence has described the King's condition as having "tertian fever (i.e., regular, intermittent fever, indicative of

Death of Saint-Louis in Tunis.

malaria) and dysentry". In 1249 again, he showed signs of dysentery: "taken up by the flows from his belly, the King has grown so thin that the bones of his spine were painfully apparent". Several other episodes of diarrhea suffered by the King have also been recorded: in 1256 in Senlis, 1259 at Fontainebleau, 1260 in Creil, 1264 at Pont-de-l'Arche. At

Tunis, on July 17, 1270, he had another attack of diarrhea and fever which led to his death. Given all the indications, he appears to have died of a combination of malaria, dysentery and also scurvy, because he also had avitaminosis characterized by gingivitis, loss of teeth, skin bleeding and bruises. Unsure about the correct procedure for embalming, his body was boiled in water and wine. His heart was placed in a monastery in Palermo (Sicily, Italy) while the bones were transported to Saint-Denis. One of his ribs is still preserved at the Notre Dame in Paris.

Phillip III the Bold (1245 – 1285), son of Louis IX has been described by historians as intellectually challenged and lacking any sense of politics. He died at the age of 40 years after an attack of dysentery accompanied by malaria, which he contracted when he accompanied his father to the siege of Tunis.

Dysentery in the armies

The armies in the field have always been prey to infectious diseases, be it typhus or typhoid fever. Severe epidemics of dysentery among the armies have also changed the course of History.

Battle of Agincourt

Battle of Agincourt, by Martial of Paris (National Library, Paris, 1448).

On the 25th October 1415, during this famous battle, 6000 archers and infantry in the army of Henry V of England were badly hit by dysentery, to the point that several of them fought without their pants on! During this battle, which is known as one of the bloodiest in the Middle Ages, the French army comprising of 50,000 men, including a large number of armored cavaliers mounted on heavily caparisoned horses, lost 10,000 soldiers and 15,000 were taken prisoners. Many of the latter were massacred as they outnumbered their guards. One consolation for the French was that dysentery continued to take its toll on the English army, which returned home with only quarter of its initial force.

Battle of Valmy

The Prussian troops of Duke of Brunswick faced the French army near Valmy-en-Champagne on the 20th September 1792. There were 47,000 French against 34,000 Prussians, who were considered to be the best army in Europe at that time. The battle itself was not so intense as the Prussians fought rather gently and curiously enough, although the number of dead did not reach even 500, the great Prussian army retreated. Turned out that the Prussians were suffering from dysentry, as evident from the latrines full of blood that they left behind. The habitants and farmers from the region called this event "the Prussian leakage" or "Prussian runnings". The disease affected the Prussian army so badly, that it barely had half of its men by the time it returned home. What a great battle!

Nevertheless, the retreating Prussian army caused enough harm, as it was followed in its wake by outbreaks of dysentery in several towns and villages, which eventually contaminated the pursuing French soldiers. For instance, at Longwy, which was held by the Prussian army until 22 October, the streets were littered

The battle of Valmy, by A. Hugo in "Histoire des armées françaises de terre et de mer de 1792 – 1783 », volume I, Delloye, Paris, 1835.

with corpses of dead soldiers. In August, Verdun, too was severely affected by the disease. At Verdun, "from every house, they threw out filth of all kinds, human and animal excreta, debris, plant waste, all of which got mixed with the mud, which would liquefy and putrefy in the rain. A foul odor emanated from it and when a carriage passed through it, one often saw people on the street stricken with spasms of nausea and vomiting or even asphyxiating".

Napoleon's army

It is estimated that one out of every ten soldiers died of dysentery during the expeditions of Napoleon in Egypt. A year later, a majority of the remaining troops was lost to the plague.

Crimean War (1854 - 1856)

This war took place between Tsar Nicolas I pitted against the French, led by Napoleon III, allied with the British government of Queen Victoria, who fought to protect the Ottoman empire. Between May and September 1855 there were 9,000 cases of dysentery and 1,478 deaths.

English sweating sickness

On the death of King Edward IV of England in 1483, his brother, Richard of Gloucester murdered the King's children in the Tower of London, and crowned himself to become King Richard III. This despicable act generated a lasting hatred from the Lancasterians and their leader Henry Tudor, Count of Richmond. There was a clash between these two enemies in Bosworth on the 22 August, 1485. Forced to fight on his feet, it is here that Richard III made his famous statement "My kingdom for a horse" before being killed. Henry of Richmond became Henry VII of England at a time when a new disease appeared in Wales. Soon after the battle, Henry went to London on 28th August. By the 19th September, the first cases of the English sweating sickness appeared in London. Shortly thereafter, the Mayor of London, his successor and several dignitaries of the Council were taken ill with this new disease, which rapidly spread all over the whole country.

Battle of Bosworth.

The attacks struck as quick as a lightening and the victims succumbed within two hours. If after 24 hours, death had not done its job, they were considered to be out of danger. This new disease was characterized by shivering, delirium, intense thirst, profuse sweating that gave out a foul odor and respiratory disorders. As it was first limited to England, it was called "English sweate" or "Sudor Anglicus". In contrast to other epidemics, this disease preferably attacked strong, healthy, young men (15 – 49 years old) while sparing the weak, the old and the children. Moreover, it did not discriminate between the rich, noble merchants and the general population. Thus, Prince Arthur, eldest brother of King Henry VIII died of this sickness at a young age, just like Henry Brandon, Duke of Suffolk and his brother Charles. Having watched the death of his secretary, the wise Ammonius, Henry VIII fled from London and went from town to town in the hope of escaping the epidemic. He left his wife, Anne Boleyn in Surrey, where she and her father were affected by the sweating sickness but survived. The ambassador of France, Jean du Bellay has recounted how he arrived at a reception in Canterbury House where eighteen people, including guests and servants, died within a few hours.

The infection persisted during the summer months and disappeared as the winter approached. It reappeared in the form of epidemics in 1508, 1517, 1528 and 1551. The sweating sickness never achieved the mortality rates comparable to those of influenza or plague, but the rapidity with which the disease progressed and ended in death within few hours, was impressive.

During the fourth epidemic, that of 1528, the disease travelled from England to Hamburg and took 1,000 victims in less than four weeks. It passed through Pomerania, then Prussia and Silesia. At the same time, it was introduced in Denmark, Switzerland, Norway, Russia and Poland. Later it gained ground in the western and southern regions of Germany, especially Bavaria. One of the most severely hit German towns was Augsburg, which had 15,000 sick and 800 dead in the first 100 days. The disease then spread to the Netherlands and caused the death of several thousand per day right at the start.

The epidemic of 1528 had one of the most interesting outcomes. Two of the greatest reformists from the Germany-speaking countries, Luther and Zwingli disagreed on several issues. Finally, in order to reconcile the two, Phillip of Hesse, invited them both to his city for discussions. This was to become famous as the Marburg Colloquium. For days, the two adversaries and their respective staff tried to convince each other. After having discussed long and hard, they seemed set to reach an agreement, when

The meeting of Luther and Zwingli.

the town was attacked by the sweating sickness. In the ensuing panic, the two rivals left without reaching an agreement, thus sealing the failure of unification of these great reformist sects. One can imagine that had unification taken place, it would have had a great impact in political and religious circles. The entire evolution of Europe and indeed of the World could have been changed.

After the last epidemic that ended in the winter of 1551, the disease never reappeared and the causative agent was never identified. In 1994, some analogies to the English sweating sickness and respiratory syndromes were induced by the Hantavirus, a class of viruses transmitted by rodents.

Typhoid

The causative agent of this disease is a bacterium, *Salmonella* Typhi.

War and famine were conducive to the spread of this disease. This was particularly true during the Napoleonic Wars, the American Civil War and the Boer War (1899 – 1902). During the latter, the British Army lost more than 8,000 soldiers to typhoid, more than those lost in fighting. During the WWI there were about 125,000 typhoid patients in the French army, of which 15,000 died of it. During the World War II (WWII), 58,000 cases were reported in the civilian population with a 10 – 25% lethality rate depending on the region.

Cholera

This is one of the oldest known disease. It originated in India and still rages in the Ganges Valley and other heavily populated areas. In ancient times, the inhabitants of this region had little communication with the outside world and so for a long time it used to be called "Indian cholera". It has been described in Athens in 400 BC, by Hippocrates. In Europe it appeared only in 1830.

Nevertheless, cholera affected all the great explorers and conquerors who passed through India. For example, an officer in the army of Alexander II the Great died of cholera, as is seen

Boer soldiers.

from the engraving on his tomb (331 BC). Descriptions of symptoms resembling those of cholera have been reported from the time of Buddha (5th century BC), Hippocrates (460 – 377 BC) and Celsus (Celsus Aulus Aurelius Cornelius, 53 BC-7 AD). Great navigators like Vasco da Gama and Magellan (late 15th and early 16th centuries) have described cases of "kholera" among of their crew.

Since 1817, six pandemics occurred, the last one ending in 1923. Leaving its origin, the disease travelled widely as a result of the developing commerce and trade links as well as wars. It spread to other Asian countries, Phillipines, the islands of Reunion and Mauritius. It travelled with the English troops to Iran in 1822, then to Syria and Arabia. In ancient Persia, it caused a large number of deaths, much to the advantage of the Russian army who conquered large regions of Iran. But the Russians were contaminated in their turn and died by the tens of thousands. Moreover, they took it to Moscow in 1830. Soon other countries were affected like Poland (by the Russian warships along the shores of the Baltic), Germany (by the merchant navy going up to Oder), England and France in 1831.

In the 19th century, many considered cholera to be a divine punishment meted out to those who had sinned and abused alcohol, food and sex. The clergy would say that the sinners died because their vital force was weakened and diminished by their bad deeds.

Cholera and the Crimean War (1854 - 1856)

In 1853 cholera appeared in a few different places in France and the following year it engulfed the entire country, especially the South. French troops who set sail in Toulon and Marseilles had already been contaminated. Some of the very sick ones were off loaded in Malta and some others in Piraeus. On their arrival in Gallipoli, 13 soldiers were showing signs of cholera and very soon others followed suit. Cases appeared sporadically with the movements of the French troops, like in Negara, Varna and Adrianople. During the siege of Varna, the English got contaminated in their turn. Between July and August 1854, the French lost 5,100 men out of a total of 55,000.

Scene from the Crimean War: Charge of the Light Brigade. Richard Caton Woodville (1825 – 1855).

The loses were equally serious among the English and the Piedmontese. Cholera spread far and wide along the sites of battles through Turkey, around the Black Sea, in Greece, Smyrna, along the coast of Dardanelles, Constantinople, Odessa and along the banks of the Danube. In addition to cholera, the troops also suffered from scurvy and dysentery. Thus, of the 97,000 British soldiers present on the battlefield, 2,700 died in combat, 1,800 from war injuries and 17,600 from dysentery, typhoid and cholera. On the whole, the outcome of the war was disastrous as it showed that one out of every five soldiers died from an infectious disease.

Cholera and the Paris Commune

In 1871, during the short-lived government of the Paris Commune, the Parisians obtained their drinking water from the Seine, the Bievre, canal of Ourcq and from the 30,000 wells in the Capital. All these sources were contaminated with fecal matter containing cholera bacillus from infected individuals!

It was the same story during the American Civil War: the Yankees lost 93,443 men in combat or due to injuries and almost twice as many or 186,216 to infections. In 1994, there was a huge outbreak of cholera at Goma in Zaire, which killed 50,000 people out of a population of 500,000 Hutu refugees who had escaped the rebels of Tutsi. After going through nearly a decade without

Cholera in Paris in 1831, by Gustav Richard Steinbrecher (1828 – 97), engraving. B/W photo by German School (19th century), Private collection/The Bridgeman Art Library, Germany.

cholera, Angola saw one of its worst epidemics during the first five months of 2006, when 43,000 people were infected and 1600 died.

Influenza

This "ordinary" infection, has been the leading cause of death for centuries. Since 1510, there were more than fifteen pandemics, that is to say, the entire planet was affected. The famous Spanish flu that hit in 1918, gave a foretaste of the consequences of a large scale microbial attack. With 30 – 40 millions victims worldwide, this epidemic caused the death of more people, civilians and military included, than the four years of WWI. The virus of 1918 displayed a virulence and a mortality estimated to be 25 times higher than the usual flu virus. It was quite inappropriately called "Spanish flu" when it was in fact American. When it first appeared in 1918, the French and German press were under a draconian censorship. Anything that might damage the morale of the soldiers was banned. In Spain, a neutral country, about one third of the inhabitants of Madrid and the King Alphonso XIII were affected. The press being free, the disease was widely discussed in the newspapers and thus came to be known as the "Spanish flu". The subsequent pandemics, "Asian flu" of 1957, or "Hong Kong flu" of 1969 caused one million and 200,000 casualties, respectively.

The potential impact of this kind of warfare, illustrated by the impact of the Spanish flu was so impressive that the in 1925 during the adoption of the Geneva protocol, the politicians of that time extended the ban on use of chemical weapons to include biological agents. Today, with the progress made in molecular biology, the generation of a more virulent and deadlier form of the influenza virus may not be a mere figment of imagination!

Spanish Flu and World War I

Seattle Police wearing masks designed by the Red Cross during the influenza epidemic (December 1918).

The first epidemic of flu was seen in the US. One morning, in the March of 1918, at Fort Riley, Kansas, a cook by the name of Albert Gitchell presented at the clinic with high fever, sore throat and generalized body ache. A couple of minutes later, another soldier appeared with the same symptoms. Very soon the number reached 107 cases and by the end of the week, 522 soldiers were affected. The epidemic at Fort Riley went on for five weeks, infected a total of 1127 soldiers and killed 46. In May, the 89th and the 92nd divisions left for France and soon after their arrival, one sees signs of flu appearing among the French. In October of 1918, 70,000 men from the American Expeditionary Force in France presented symptoms of flu and General Pershing was forced to ask for reinforcements. In all, a third of the patients died of flu and in some of the small units the death toll was almost 80%. The consequences were so grave that on 27th September 1918, a notice was sent by General Crowder postponing the incorporation of 14,200 new recruits initially planned for 7th – 11th October to the 23rd October. Due to the overcrowding of the recruits on boats travelling to France, the disease spread easily among the crew and at times the mortality reached 20%.

Officers of the American Expeditionary Force and Mission Baker.

The President Woodrow Wilson had to take a delicate decision about cancelling or not, the reinforcements as it appeared to cost several American lives. At the same time the German troops were weakened by the four years of conflict and the specter of defeat was slowly creeping into their minds. On 8th October, General March, who was in favor of continuing the war, went to Washington to meet President Wilson to persuade him to continue to send fresh troops to Europe, emphasizing that "every soldier who has died (of influenza) just as surely played his part, as his comrade who died on the front in France". Finally President Wilson was persuaded by his arguments and continued to send out troops. A little more than a month later, armistice was signed and the War ended.

Thomas Woodrow Wilson (1856 – 1924), President of the United States of America.

The Spanish flu also played an important role in the German camp. After the retreat of the Russians in the Spring of 1918, the Germans decided to relocate all their forces to the western front, which included a million soldiers, 37 infantry divisions and 3,000 canons. On 21st March 1918, they launched their last forces in the battle aiming to gain victory or at least to discourage the Allies and be in a favorable position to negotiate the terms of a cease-fire. They assembled large numbers of troops from all over the front to break the Allied lines. Their advances were finally halted at Marne, a few kilometers from Paris. On 18th July, the offensive of Foch with a flank attack

I WANT YOU
FOR U.S. ARMY

A poster to recruit soldiers into the US army during the World Wars.

led by Mangin from the Forest of Villers-Cotterets, together restrained Ludendorff, forcing him to retreat and thus opened the path to victory.

According to most historians, General Erich von Luddendorf was obviously close to a great victory that would have completely changed the course of the War. However, curiously enough, he did not exploit this highly favorable situation, but gave his troops one month to recover. This allowed the French, British and Americans to recuperate. From several reports of that time, it appears that one of the chief reasons for his failure to score was the flu epidemic that was causing heavy losses in his troops. This fact is confirmed by General Ludendorff himself in his memoirs published in 1919 after taking refuge in Sweden.

He blames his failure at the battle of Marne to the fact that his troops were decimated by the flu epidemic, which the soldiers called "Blitzkatarrh". He writes that "it was a painful task every morning to listen to the litany of number of cases of flu and the complaints about the fatigue in the troops".

Erich von Ludendorff (1865 – 1937).

Another indirect consequence of the flu epidemic resulted in placing Adolf Hitler in a position of power, a few years later. During the negotiations of the Armistice, President Woodrow Wilson of America contracted influenza during his stay in Paris. He was rendered so sick and weak that he accepted some of the conditions being defended by President Georges Clemenceau of France that were in fact harsh vis-à-vis Germany. In particular, this included declaring full responsibility of the conflict on Germany, demilitarization of the Rhineland, permitting the operation of the Saar mines by the French and the return of Alsace and Lorraine to France. In addition, the German Air Force could no longer exist and its military forces would be limited to 100,000 men. The German colonies in Africa would be redistributed to the allies. These conditions being extremely hard on Germany, destroyed the German economy and created fertile grounds for generating a spirit of nationalism. It resulted in the rise of the Nazi Party to power, a political chaos, and subsequently in WWII.

Georges Clemenceau (1841 – 1929).

From the Dark Ages to Modern Times

Potato mildew and President John F. Kennedy

About 150 years ago, a microscopic fungus, *Phytophthora infestans*, destroyed potato crops in Ireland and brought on the greatest European social catastrophe of the nineteenth century.

At that time, the English regarded the Irish with great disdain, and this was to have its repercussions several decades after the Great Famine. The anti-Irish racism ran strong, as evident from the writings of English authors who described Ireland as "a small island at the extreme corner of Europe, that had not yet come out of the Stone Age"! Benjamin Disraeli, the Prime Minister in the reign of Queen Victoria said, "The Irish hate our order, our civilization, our spirit of enterprise". The very Victorian English historian, Charles Kingsley recorded his observations during the Great Famine, "I am haunted by the human chimpanzees that I saw all along the 100 miles of this horrible country. I don't believe they are our fault. I think there are many more who are better fed and housed under our rule than ever before".

Ireland of the early 19th century was one of the most densely populated countries in Europe and also one of the poorest. According to an 1830 estimate, it had about 3 million needy people. It had to do with laborers who rented their land to rich farmers, who in turn were dependent on owners, most of whom lived in England. Thus, in 1842, around 6 million sterling pounds changed hands from Ireland to England. At the same time, about a third of the cultivated land was under potato plantations, as this was the staple food of this country. Potatoes were a substitute for bread, eaten with fish and butter.

The fungus, *Phytopthora infestans*, that probably originated in Peru, is the causative agent of potato late blight or potato mildew. The fungus was first encountered in the US in the summer of 1843. It was transported by boat to Belgium and by the summer of 1845, it had invaded the region Flanders, in Belgium, Normandy in France, Holland and southern England. The English Prime Minister, Robert Peel considered it to be "a calamity for the poor". In Ireland, the epidemic was first noticed in the Botanical Garden in Dublin in August and just about a week later, the harvest of Fermanagh was destroyed for the most part. The same year, roughly half of the potatoes from the west of Ireland were either destroyed or were not fit for consumption. The following year, a heat wave followed by heavy rains created conditions that were ideal for the growth of the fungus and led to the loss of approximately 90% of the harvest.

The great Famine of 1845-1849 in Ireland was a consequence of a mildew that destroyed potato plantations

The winter of 1846 – 47 was particularly severe. In early 1847, called "Black 47", there were reports of corpses being eaten by street dogs. In that year, more than a million died of hunger and diseases like typhus or cholera. Malnutrition due to the famine had rendered the population so weak and prone to infections, that between 1845 – 50, nearly 1.5 million out of the 8 million habitants were estimated to have died from diseases like typhus, cholera, dysentery or repeated fevers.

Immigrants leaving Ireland, by Henry Doyle (1868).

Hence, one finds that in the four years following Black 47, two million Irish people left their country for good, and migrated to other English-speaking countries like the US, Canada and Australia, as well as to the major industrialized cities in England like Liverpool and Manchester.

This massive migration had a considerable social impact. For instance, in the US today, one person in every ten has Irish roots. In 1914, there were 5 million Irish-Americans. These immigrants were concentrated in cities such as Boston or New York, where they had first landed. They soon began to occupy strategic positions in various sectors: railways, mining, civil engineering, justice and politics. Prior to 1840, the white population in the US were mostly Protestants who had come from England, Holland, Germany or Scandinavia. There were fewer Catholics, those from France, Spain, Italy, Switzerland or Germany. With the arrival of the Irish in large numbers, Catholic ghettos were created and in some cities reached 30% of the population.

John Fitzgerald Kennedy, 35th President of United States of America.

Their political clout was to grow and have important consequences. The migrants, poorest of the poor, swelled the ranks of the Democratic Party, which holds true even today. An anti-imperialist and isolationist sentiment was reinforced by the Irish during the 20th century because of their hatred for the English. It is almost certain that the anti-English lobbying played a role in delaying America from participating in the two World Wars. The strong anti-English stand of the Irish came to the fore again during the Civil War. The Irish community voted for the northern Republicans not because they were abolitionists, but because they were fiercely opposed to the English, who were represented by the "WASP" or White Anglo-Saxon Protestants, mostly from the southern states.

Among the first migrants, was an Irish family of Counts from Wexford and Kerry, named the Kennedy's. They were one of the first "Boat people" of modern times. One of their descendants was

John Fitzgerald Kennedy (1917-1963) who went on to become the first Catholic and the 35th President of United States of America, holding office between January 20, 1961 - November 22, 1963.

Potato mildew and WWI

Another interesting consequence of the infection by *Phytopthora infestans* is seen during the WWI and concerns Germany. A terrible famine afflicted this country in 1916 – 17 taking 700,000 victims, essentially civilians, children and senior citizens who had been weakened. During the War, copper was used to manufacture shells, shrapnel and electrical wires and hence was less and less available for making the boiling slops that were used as a preventive against potato blight. Moreover, the potatoes and cereals that were edible were provided to the soldiers at the frontlines, while the civilians left behind had to subsist on turnips and whatever other foods they managed to find. In the autumn of 1916, the German High Command decided to concentrate all their forces on the western front, the eastern front being weakened. However, the desire for victory had dwindled considerably among the German soldiers who were aware of the plight of their families back home, facing the famine. We have also seen earlier that the German High Command was unable to seize a victory due the influenza outbreak on the western front. Finally, the Germans retreated and maintained their position on the Hindenburg Line. The arrival of the American troops in support of the French and the English was to be the final blow for the Germans.

Tetanus

The symptoms of this disease are known since the Dark Ages. According to Hippocrates, a wound however small should be considered as a possible cause for tetanus. The disease was common during the conflicts. Tetanus has appeared very frequently during conflicts such as, the War of the Austrian succession (1740 – 1748), or those of the Holy Empire, or the Franco-Prussian War of 1870, and particularly during the WWI, when it flared up. In 1803, during the campaign of Egypt and Syria, Dominique Larrey, Chief surgeon of the Napoleonic Army noted that "this disease starts with a generalized malaise and uneasiness that seizes the wounded...it is accompanied by acute pain that is worsened by contact with air and even the lightest external objects; the entire limb becomes painful. The irritation rapidly spreads from the muscles surrounding the wound to distant ones, which then contract; or all of a sudden it may run through the body and stay concentrated in the muscles of the throat and jaws".

The causative bacteria *Clostidium tetani* enter the body through a wound and cause horrible suffering. Death occurs due a respiratory failure as the lungs can no longer function properly. The normal and essential habitat of these bacteria is the soil and the intestinal tracts of man and animals, especially herbivores and it can exist as spores (resistant forms). Presence of *Clostidium tetani* in the gut of horses is closely associated with contamination of the soil by excrements of these animals.

Soldier suffering from opisthotonus caused by tetanus, on a battlefield, by Sir Charles Bell (1809).

Regions with a high density of equine population are known to be particularly "tetanigenic", for example stables and race courses in general, but may also concern entire regions, like the Varreddes in Seine-et-Marne in France. During WWI, a large number of horses were used in this region to transport guns, which made the land "tetanigenic". Thus, after 1915, many a soldier who came in direct contact with the bacterial spores in this region had to be buried in the trenches. These soldiers were the ones who paid heavily for this infection. The highest incidences of tetanus after wounding were observed during the battles at Marne and Aisne, whereas there were fewer cases during the battle of Ypres. This corresponds no doubt to the quantity of the spores present in the soil. The rich soil of Aisne was so notorious for tetanus that several years before the War, farm horses were already protected by the tetanus anti-toxin. Tetanus was comparatively rare on the other fronts.

Serratia, "the red bacteria"

Serratia is a peculiar bacterium that secretes a red pigment or "blood". Prior to the identification of this bacterium in the late nineteenth century, the pigment gave rise to all sorts of speculations, including the famous "bloody wafers" stained by the blood of Christ.

The first encounters

Pythagoras. Detail of School of Athens, by Rafaello Sanzio (1509).

The first observations about the ability of these bacteria to produce a bright red pigment in media that are rich in starch such as beans, bread, polenta or wafers, date back to the 6th century BC. Pythagoras (560 – 500 BC) is one the most mysterious figures of ancient Greece. Never having written anything himself, all his teachings are known from the writings of his disciples. Very soon he became a legend. They called him the son of Apollo or Hermes, from whom he had received the power to remember his past lives. However, this does not refute the existence of this philosopher.

Today Pythagoras is mostly known as a mathematician, astronomer and geometrician, whereas in the Dark Ages he was considered

From the Dark Ages to Modern Times

rather as a philosopher. He believed strongly in reincarnation of the soul and was a vegetarian. In his way of living, he was equally against drinking as eating beans! Probably because he believed that the soul of his parents was inside the beans. During his stay in Egypt, as a result of this theory, beans were banned among the priests although they were part of the traditional Egyptian cuisine. Incidently, beans were also banned by the followers of the prophet, Zarathustra or the Zoroaster, a native of Azerbaijan, who lived around the 7th – 6th centuries BC, and whose life was nothing but a succession of miracles.

In the year 170, Lucian of Samsat (today Samsat is in Turkey) (ca. 125-ca. 192) wrote in his Viarum Auctio, a dialogue between a disciple of Pythagoras and Agorastes, who asked him why he was allowed to eat all vegetables except the beans. The answer was: "beans are sacred. Take a green bean and remove the skin. Boil it and expose it to moonlight for a few nights and you will see the blood". The "blood" was probably from the red pigments produced by the bacteria, *Serratia marcescens*.

Siege of Tyre by Alexander the Great (332 BC)

A more precise reference is found during the siege of the Phoenician city of Tyre (today in Lebanon) by Alexander the Great in 332 BC. At that time Tyre was the main commercial core of the Middle East and a major naval base. The Greek historian, Diodorus of Sicily, who lived in the 1st century BC has described the siege and recorded what are probably the first cases of infections of *Serratia marcescens* in history.

In 332 BC, Alexander followed his plan of systematic encircling of the eastern Mediterranean with aim of gaining control over the region. He occupied Syria, then Phoenicia. After seven months of siege, the Macedonian soldiers were getting fatigued, when a soldier noticed "drops of blood" on a slice of bread: "after the rations were distributed to the Macedonians, the bread appeared to bleed as we broke it". Similarly, the historian Quinte Curtius (1st century AD.) writes that "in the Macedonian camp, soldiers noticed that the bread dripped drops of blood … ".

A soothsayer named Aristander interpreted the event as a good omen "If the droplets of blood had come from outside, the Macedonians would have been in danger, but as the liquid comes from within, it was a bad omen foretelling that Tyre would be destroyed ". Indeed, Tyre was taken in 332 and this city that had made the mistake of resisting Alexander was razed to ground: 8,000 inhabitants were exterminated and 30,000 sold as slaves.

Bread and bloodied wafers

In 1169, at Alsen (Denmark), a priest claimed to have seen a wafer soaked in blood. On hearing this the Father Superior predicted a "blood bath" for the Christians. Two weeks later, an army consisting of pagan Slavs descended on the region, destroyed churches, sent people into slavery and killed those who resisted.

Later, other cases of bloody wafers were observed in Italy, Germany, Spain, France, Belgium, Poland and Austria. These events were almost always an opportunity for the gullible masses of the Middle Ages, to accuse the Jews of uttering insults against the Sacred Host. It is estimated that more than 10,000 unfortunate Jews lost their lives to these superstitions.

In 1896, Scheurlen noted that "dieser Saprophyt mehr Menschen umgebracht has als mancher pathogene Bacillus", meaning that this saprophyte killed more humans than many of the bacterial pathogens".

The US Army experience

From the 1950s to the middle of the 1960s, *Serratia marcescens* was used to simulate biological warfare by the US army, at several places like the subways in New York, San Francisco Bay Area, and in the cities of Key West in Florida and Fort McClellan in Alabama. Between 20 – 27 September 1950, a US Naval ship, cruised through the San Francisco Bay Area spraying aerosols of *Serratia marcescens* and *Bacillus globigii* over an area of 300 km. The bacterial suspensions were mixed with fluorescent particles of zinc sulphate and cadmium to allow an easier detection and determination of the impact. According to the military estimates, each of the 800,000 inhabitants received a dose of around 5,000 particles. Although these microbes were supposed to be completely harmless, two weeks after their release, on October 11, 1950, Edward J. Nevin was admitted to Stanford Hospital in San Francisco with fever and signs of pneumonia. At the same time, ten other patients with similar symptoms were also admitted to the hospital. Edward J. Nevin died three weeks later due to a *Serratia marcescens* endocarditis, but the other patients recovered. According to the retrospective investigations carried out by the Centers for Disease Controls (CDC), Atlanta, these cases were not related to the U.S. Army operation. This has however been questioned because after such simulations were carried out at Fort McClellan, Alabama in 1952, the number of cases of pneumonia in the county doubled (New York Times, March 13, 1977).

On 17th September 1975, William E. Colby, who was the Director of the Central Intelligence Agency (CIA) from 1973 – 76, deposed before the Senate and informed the non-believers about the Project "M.K. Naomi" concerned with the experimental simulations of biological attacks that were carried out in 1966 in New York subways. The bacteria were released from a small balloon placed in a running train. The aim was to follow the movement of the aerosol released on the lines located between the 6th and 8th Avenues during the rush hours. Russell Baker (New York Times, 23rd September, 1975) humorously asked for a compensation from the US Army to the New Yorkers for serving as guinea pigs.

William E. Colby, Director, Central Intelligence Agency (CIA).

From the Dark Ages to Modern Times

All of the 239 trials carried out by the US Army, using *Serratia marcescens* were disclosed on 22 December 1976, including an "attack" against the White House in 1962 which revealed defects in the air filters and other weaknesses in the building in terms of facing a biological attack.

Although to the best of their knowledge, no cases of infection or death were caused by these trials, the Pentagon finally abandoned them in 1970 considering them to be too risky to use on "friendly populations".

These experiments were permitted in those days because the "red bacteria" were mere laboratory curiosities and had rarely been implicated in the any infectious process. The situation has changed considerably since then because the bacteria have mutated, giving rise to clones of non-pigmented strains of *Serratia*, which are now considered as one of the leading causes of hospital-acquired infections.

Anthrax

Anthrax is caused by the bacterium, *Bacillus anthracis*. There are three forms of the disease depending on the mode of infection: a cutaneous form which causes black sores on the skin, a gastrointestinal form and a pulmonary form, which if untreated, is usually lethal.

Anthrax in mythology

The Plagues of Egypt

The origin of anthrax is contemporary with pastoralism and it has accompanied civilization ever since. The first descriptions of this disease are found in the Bible, dating back to 1500 BC. After a famine in the land of Canaan, God had gone through Israel to Egypt. As time passed, one of the Pharaohs, Sethi I, father of Ramses I, reduced the Israelis to slavery. In spite of this, these descendants of Abraham did not stop increasing in number, and the Pharoah became worried about seeing

The 5th Plague of Egypt, by William Turner (British Museum, London, 1806 – 1810).

them rise in revolt one day, which would mean losing the cheap labor and then the business. So he passed an order that all the newborn boys be killed. However, one child escaped death: little Moses. His mother had hidden him in a wicker basket, and set it afloat on the Nile. Later, when he grew up, he tried to flee from Egypt along with his people, to escape the army of the Pharaoh. Due to the latter's stubbornness, God is said to have proclaimed ten successive calamities on Egypt: known as the "ten plagues" of Egypt. In the fifth plague, sheep, cattle and horses were ravaged by an infection, the symptoms of which correspond to anthrax, indicating that the disease may have been common at that time. On leaving Egypt, Moses and his people reached the Red Sea, where God opened it only for them to pass. The Egyptian army thus could not prevent them from reaching the promised land.

Hercules and Deianeira

Abduction of Deianeira, by Guido Reni (Louvre Museum, Paris, 1620 – 1621).

Legend has it that Hercules who had come as a guest of the King Dexamenis, took the opportunity to deflower his daughter Deianeira and later married her. One day, Deianeira requested the centaur Nessus to take Hercules and her across the river Euhenus. Nessus agreed, but in midstream tried to rape her. Hercules killed Nessus with his arrow. The latter knew that the arrow had been dipped in the gall of Hydra of Lerna and contained a potent venom. Collecting the blood oozing from his own wounds, he gave it to Deianeira telling her that it carried the potion of fidelity: if she ever wanted to prevent her fickle husband from abandoning her, it would be enough to soak his clothes with the blood. Deianeira believed him and carefully guarded this secret. Years later, when Deianeira learnt that Hercules was bringing back Iola, a young prisoner of extraordinary beauty, she feared her marriage would break.

Therefore, remembering the advice of Nessus she sends Hercules clothes soaked in the centaur's blood. When Hercules wears the clothes, he immediately feels as if he has been set on fire. On diving into a river to put out the fire, he feels bigger flames coming out of his body. When he tries to tear the clothes off, pieces of his flesh come off. Finally, he is driven to end his life. As for Deianeira, when she hears of the tale, she too kills herself. The suffering of Hercules in this story corresponds perfectly to symptoms of anthrax. Yet another proof that the disease was already well known in the Dark Ages.

"Black Bane"

In the late Middle Ages, anthrax spread all over the European continent. In Great Britain, it was called the "Black bane" or the "Black Death epidemic". In the 17th and 18th centuries, it caused significant damages in Europe. It is said to have resulted in 60,000 deaths and the correlation between the disease and the trade of sheep wool, that contains the spores of the bacteria has been clearly established. There have been more recent epizootic outbreaks like the one that killed more than a million sheep in Iran in 1945. It was since the WWI, that the bacillus of anthrax was used as a biological weapon and achieved irrefutable success throughout the 20th century as will be described in the last part of this book.

6 The Modern Age

Pasteur, the "innovator" of biological weapons

Louis Pasteur was the first to have the idea of fighting one living organism with the help of another. He had proposed a plan to control the rabbit population in New South Wales in Australia by contaminating the rabbit feed with a pathogenic bacteria that causes chicken cholera. It was "...a poison for those endowed with life...". Although the cultures were bought by the Australian government the project was abandoned.

Nevertheless, the model was used in Champagne, in 1887. Madame Pommery, wanted to rid her vineyards of rabbits because their burrowing caused loose stones to fall on the bottles of Champagne stored in her cellars. The control measures employed was to spike the rabbits' meals with fresh cultures of the fowl cholera bacillus. This was the first manmade epizootic epidemic. The results were rapid and spectacular. There were no rabbits seen alive and in a span of three days, 32 dead ones were collected. Since most of the rabbits had probably died in their burrows, the actual number of dead must have been much greater than that reported.

Madame Pommery.

This method was also emulated by a forest owner in 1953. In order to reduce the number of rabbits that were destroying his property, he inoculated two of them with Myxoma virus. The results exceeded all expectations as the virus spread uncontrollably throughout this region, and then all over Europe, killing wild rabbits as well as those on farms. The livestock industry suffered damages and France was brought before the Court of International Justice at The Hague!

Use of bacteriological weapons on a "small-scale"

With the progress made in microbiology during the 20th century attempts were made at using microbes and their toxins for hostile purpose, and some achieved significant success.

Henri Girard

Son of a pharmacist, Henri Girard was expelled from school on charges of theft and other irregularities. Later, as an insurance broker, he found an ingenious way to defraud his parents, neighbors, friends and customers. When he offered life insurance, he would fool the client into signing two policies, of which one would be either in the name of one of his mistress or in his own name. An example of his modus operandi is seen in the case of Helen Monin.

April 30, 1918: a young Parisian called Helen Monin was invited to lunch at the place of her friend, Henri Girard, an insurance broker. After an excellent meal, she headed for the subway station at the Place d'Italie, when she was overcome with a sudden malaise. A severe vertigo and nervous tremors nailed her to the spot. A passerby helped her into the first house he could find, where a concierge asked her to lie down while he rang for a doctor. Helen was writhing in pain as the doctor examined her and despaired, "This is a rare case. I think it is typhus". Helen died an agonizing death while the doctor watched helplessly. Little did he know that the lady had eaten heartily of a dish of mushrooms carefully prepared by Henri Girard, a fine connoisseur. A perfect crime, or almost.

The truth was to come out later. Helen Monin had drawn four life insurance policies for a huge sum, at the Phoenix Company, where Girard worked. In fact, these contracts had been signed in person by one of Girard's mistresses, Jeanne Drounin, under the guise of Helen Monin. On the death of the latter, Jeanne came forward to collect the money as per the policy and plan. This reckless mistake proved fatal. An employee of the insurance company recognized Jeanne as the late "Helen" who had signed the policies and was taken aback to see a dead person come to collect insurance money! The police was alerted and Girard was arrested.

It later came to light that Henri Girard was cultivating a host of different bacterial cultures at home, and would serve the bacteria to his guests. From laboratory sources, he had obtained cultures of *Salmonella* Typhi, the causative agent of typhoid, *Bacillus anthracis*, causative agent of anthrax and the Koch bacillus that causes tuberculosis. Jailed in the Fresnes prison, he died of tuberculosis in the infirmary in 1921. In the lining of his coat they found a vial containing a culture of the tubercle bacillus.

The Pandey brothers

1929: a man named Pandey dies in Pakur, India, living behind two sons. In 1932, the elder son makes the first attempt to kill his brother by simply scratching his skin with a pair of reading glasses while pretending to help him put them on, somewhat forcefully. The younger brother gets tetanus but is saved by the timely application of the antiserum. A year later, the two brothers are at the Kolkata (earlier, Calcutta) station when the younger one is bitten on the arm. This time, he dies quickly after a high fever. However, a blood culture reveals the presence of *Yersinia pestis*, the agent of plague, although the disease no longer existed in Kolkata. Investigators later find out that the elder brother had made several unsuccessful attempts to procure the plague bacillus from Haffkine Institute, Mumbai (earlier, Bombay) and had ended up stealing it from the hospital.

Botulism and Heydrich

Botulism is a food-borne disease caused by the anaerobic bacterium, *Clostridium botulinum*. In adults it is not the bacteria that makes the person sick, but rather the toxins it secretes.

Reinhardt Heydrich,
"the torturer of Czechoslovakia".

During the WWII, the English research center at Porton Down, that had earlier developed the anthrax bomb, also developed one with the botulinum toxin. In October 1941, Paul Fildes, the director of the centre, was approached by the British Secret Service with the proposed Operation Anthropoid, a plan for assassinating Reinhardt Heydrich. Heydrich had acquired a formidable reputation as the head of the Sicherheitdienst (SD), the Nazi Security Service. He was the heir apparent designated by Fuhrer Adolph Hitler and was the Protector (Reichsprotektor) of Bohemia and Moravia in September 1941. His control was fiercely repressive, but in this region he was able to efficiently develop an armament industry that soon assumed economic importance in the Third Reich.

Operation Anthropoid was launched in December 1941. Seven supporters of the Czech resistance landed at night by parachutes near Lidice, a little Bohemian town. They carried arms and radio equipment. This included two anti-tank hand grenades no. 73, that had been personally prepared by Paul Fildes at Porton Down and had very special characteristics. Each grenade weighed a pound and its upper one-third was covered with an adhesive tape tightly bound to the open end of the device. Led by Jan Kubis and Josef Gabcik, the supporters stayed low for five months, studying the movements of Heydrich. They swung into action on May 27, 1942. There were six of them: Rela Fafek, Gabick's girlfriend, 4 men armed with revolvers and grenades, and the sixth one positioned strategically with a mirror in hand to signal the arrival of Heydrich's car at a turn near the Troja bridge on the outskirts of Prague. Rela was to precede Heydrich's car and signal if he was escorted or not. Surprisingly, for a Nazi officer of his stature, he often travelled without an armed escort. As his green Mercedes rounded the corner, Gabick was standing right in the middle of the road holding a submachine gun and immediately opened fire. Heydrich cried to his driver to accelerate but it was too late as Kubis, the other leader, lobbed one of the two grenades at the car. The grenade missed the target but the explosion ripped the car door and sent splinters and shrapnel that hit Heydrich. He jumped out of the car revolver in hand, screaming and holding his right hip. While the attackers dispersed, he staggered and collapsed and was taken bleeding heavily, to the nearest hospital in Bulovka.

The car of Heydrich after the attack.

Among the numerous injuries, a splinter of the grenade or a metallic piece from the car were found in the wall of his lungs, near the spleen. There was a fractured rib and the diaphragm had been punctured. However, the injuries did not appear alarming or life-threatening to the doctors attending him. The operation to remove the splinters and shrapnel was successful and everything went perfectly fine. The next day, the situation worsened suddenly and by evening Heydrich sank into a coma. He died June 4th, 1942. In retrospect, he had displayed symptoms that correspond to botulinum poisoning: extreme weakness, malaise, dry skin, dilated pupils, dry tongue and mouth. These were accompanied by muscle weakness, paralysis of the face, legs, arms and breathing muscles. But, according to the doctors, his death was inexplicable. The German authorities concluded that "death occurred as a consequence of lesions in the vital parenchymatous organs caused by bacteria and possibly by poisons carried into them by splinters of a bomb splinters...". Later, Fildes claimed responsibility and boasted that the killing of Heydrich was the "first notch of my pistol".

7 Towards weapons of mass destruction

During the 20th century, new steps were taken towards the use of biological weapons on a large scale. Countries all over the world have engaged in the pursuit of making these weapons of mass destruction through ingenious ways. A few examples from recorded history are given below.

German trials during the WWI

Experiments of Anton Dilger

The Germans had developed a major research program on biological weapons during WWI. The program was based on contamination of the animal feed that would be exported to the Allied troops. In 1915, Anton Dilger, a German-born American physician established a small laboratory in his hometown northwest of Washington where he produced of cultures of *Bacillus anthracis*, the causative agent of anthrax, and *Burkholderia mallei*, the causative agent of glanders. He bribed the dockers at the Baltimore port and managed to infect 3,500 mules, horses and other animals that were sent for the troops fighting on the European front. Several hundreds of soldiers were infected in their turn. This phenomenon ceased in the months following the return of Dilger to Germany at the beginning of 1916. Apparently Dilger died in Madrid on the 17 October 1918 under the name of Alberto Donde, during the pandemic of Spanish flu (see the section on "Influenza").

German legation in Bucharest

27August, 1916: a package addressed to the Bulgarian imperial legation in Bucharest, Romania, arrived at the German legation in the city. It was marked "Top Secret. Hold right side up". At this time of Germany occupation of Romania, in-coming packages went through immediate

A cavalry during the WWI.

dispatch from the German legation. But if the packages were bulky, a certain Michel Markus would bury them in the garden on the premises. In October 1916, during a search by the Romanian and US authorities, he was asked to recover this particular package. It was found to contain a small wooden box with vials containing a yellow-colored liquid and a note saying, "Enclosed are: 1 vial for horses, and 4 for cattle. Use as agreed. Each vial is sufficient for 200 pieces. As far as possible, inoculate directly into the mouth, failing which, mix with the feed". Laboratory investigations on the contents of the vials revealed presence of the bacilli of glanders (*Burkholderia mallei*) and anthrax (*Bacillus anthracis*) meant for horses and cattle respectively.

Tularemia at Stalingrad during WWII

Tularemia is due to the bacteria, *Francisella tularensis*, which are found in Europe and North America. The bacteria are very resistant to cold and notably to freezing. Small mammalian rodents such as voles, field mice, squirrels and muskrats or lagomorphs such as rabbits and hare, are susceptible to the disease and serve as reservoirs for the bacteria.

Scenes from the Battle of Stalingrad.

The siege of Stalingrad is a turning point in WWII. It is one of bloodiest battles in human history with 2 million dead, of which 1.2 million were from the Soviet camp. On the 22 June 1941, Fuhrer Adolf Hitler attacked USSR. His troops ran over the immense Russian plains with great speed and arrived at the gates of Moscow and Leningrad. Those from Wehrmacht marched south towards the oil fields of Caucasus, whereas the 6th Army of General Friedrich Paulus veered towards Stalingrad to cut the Soviet supply lines. Stalingrad was defended by the Soviet Chief of Staff, General Georgi Zhukov. The city extended over a distance of 40 km, and in the autumn of 1942, was conquered street by street at the cost of immense suffering on both sides. On 19th November, Paulus finally managed to occupy the city, but his troubles had just about begun. Zhukov unleashed a powerful counter attack. Sensing that the trap was closing in on his men, Paulus asked for permission to retreat, which was refused by Hitler. Eventually in February 1943, Paulus surrendered along with 24 of his generals and 91,000 soldiers who had survived, of which only 6,000 returned from captivity. Among the casualties, many soldiers from both sides had developed infectious diseases. A defector, Ken Alibek or Kantian Alibekov, who had worked for long for the USSR in the Biopreparat, a large centre for development of biological weapons, reported that hundreds of thousands of soldiers suffered from tularemia during the Battle of Stalingrad. He put forth the hypothesis of a purposeful contamination of the regions where fighting had taken place with the bacteria *Francisella tularensis*, causing a large number (70%) of the lung infections. However, this hypothesis is not accepted by all historians, since in the Rostov region, 12,000 cases of tularemia had been recorded in January 1942, several months before the arrival of the first Panzers. Hence this was probably a natural outbreak often seen during times of war when basic sanitary precautions and elementary hygiene are ignored and in the event of uncontrolled multiplication of rodents. Indeed, due to the proximity to the war front, farmers were unable to harvest their crops, which was thus plentifully available for the rodents that multiplied all over the place.

The surrender of Field Marshal Paulus.

Nazi Germany

While the Germans excelled at chemical weapons, they had carried out little research in biological ones. The only known attempt was conducted in 1932 in the subways in London and Paris, and consisted of spreading suspensions of *Serratia marcescens*, a non-pathogenic bacterium.

Hitler had ordered the development of biological weapons during the WWII. With the support of the Nazi regime, the scientists undertook the research but progressed rather slowly compared to those in other countries. Trials of contamination with different pathogenic agents were conducted on prisoners in concentration camps. The only known application of a pathogenic agent by the Nazis was that of contaminating a water reservoir in the northwest of Bohemia in May 1945.

Polish "typhus" epidemic

The Germans were scared of the epidemics of typhus and this fear was exploited by people from occupied Poland in a trick to evade compulsory labor. As per the rules, only those who were free from the disease were taken to Germany. The Polish population was screened for typhus based on diagnosis by the Weil-Felix reaction, a serological test (which does not require culturing of the bacteria). The presence of *Rickettsia prowazekii*, the causative agent of typhus, in the serum of infected patients results in a positive Weil-Felix reaction. Interestingly, another bacterium, *Proteus* OX19, has antigenic similarity with the agent of typhus and hence gives a false positive reaction in this test. Armed with this knowledge, two Polish physicians, Drs. Eugeniusz Lazowski and Stanislav Matulewick "vaccinated" the population by injecting the relatively harmless *Proteus* to induce a false positive reaction to typhus. A fictitious epidemic of typhus thus saved several Polish people from deportation.

Great Britain

Island of death.

Gruinard Island is a string of small islands of Hebrides, northwest of Scotland. It was the site of biological weapons testing carried out by the British scientists during the WWII. They tested different projectiles containing several millions of anthrax spores on the sheep confined to the island. The sheep died within a few days of exposure. These tests caused such a massive contamination of the soil that it took 37 years before the soil samples collected from Gruinard Island showed any significant reduction in the number of anthrax spores. The island was decontaminated in the 1980s. Today it is totally safe and the herds of sheep that were released there are healthy. The aim of these experiments remains controversial, but it seems obvious that it was to develop an arsenal of biological weapons against the Germans.

Japan

In 1928, Shiro Ishii, a brilliant and ambitious Japanese doctor managed to convince the Japanese authorities about the potential importance of biological warfare. In 1930, a Department of Immunology was created at the Army Medical School in Tokyo. In 1932 at Pin Fang, near Harbin in Manchuria, an entire research unit for biological weapons was created and was later known as the notorious Unit 731. Similarly, two other centers were opened, the unit 100 near Changchun and the Department Tama near Nanjing. This research program was directed by Shiro Ishii from 1932 – 1942, and by Kitano Misaji from 1942 – 1945. The Unit 731 included 150 buildings, 5 satellite camps and employed 3,000 scientists and technicians. The trials consisted of infecting several prisoners with cultures of pathogenic bacteria such as those causing anthrax (*Bacillus anthracis*), meningitis (*Neisseria meningitidis*), dysentery (*Shigella* spp.), cholera (*Vibrio cholerae*) and plague (*Yersinia pestis*). The experiments resulted in the death of 10,000 prisoners under horrific conditions, surpassing those prevailing in the

Shiro Ishii.

Nazi concentration camps. According to the testimonies of those who participated in these programs, at least 11 villages in China were subjected to biological attack. A total of 250,000 people were killed in all of China.

Unit 731 was evacuated in 1945, but its existence was revealed only in 1949 during the famous trial of Khabarovsky, where only a dozen people were tried. Shiro Ishii had negotiated his release from the US Army. After the

War, the American government promised him pardon in exchange for the results of his research. Thus Shiro Ishii, one of the deadliest criminals of all times, returned to Japan and lived peacefully until his death in 1959 from throat cancer. A monument was erected in his honor in his country!

Unit 731.

United States of America

The US research program on biological weapons began 1942. It included various sites for research and development, in Fort Detrick, Maryland, testing in the states of Mississippi and Utah and production of equipment at Terre Haute. They used mostly anthrax (*Bacillus anthracis*) and brucellosis (*Brucella suis*). In WWII, the results of this research led to the production of least 5,000 bombs loaded with anthrax.

After the War, several of the Japanese scientists from the erstwhile Unit 731 had to work at Camp Detrick and reveal the results of their research program in exchange for political immunity. Their ideas were implemented during the Korean War, during which fleas infested with the plague bacteria were used by the Americans. A clear correlation was noticed between the passage of US aircrafts and the sudden appearance of fleas in large numbers in the months of January and February 1952. The fleas had been released in the village of Balnam Ri of Anju on the 18th February and plague broke out on the 25th. Out of 600 inhabitants, 50 were affected and 36 died. Similarly, towards the end of March 1952, an airplane flew over the village of Kang-Sou, and soon afterwards a farmer reported a large number of fleas floating on the surface of the water in a huge jar that was placed outdoors. A few days later, he died of plague.

The 1972 convention.

The advanced techniques in the development of biological weapons by the Japanese scientists covered aspects of cultivation, concentration, storage and aerosolization of the microbes. In parallel, they conducted a program concerned with defense measures, in case of a biological attack, which included developing vaccines, antisera and antibiotics. In 1955, they carried out their first trial on military and civilian volunteers.

Eventually, the Americans realized that the military use of such weapons was not effective due to the period of latency between the release of an agent and the appearance of the symptoms of the disease. Moreover, it is impossible to limit the zone of infection. On 25th November 1969, President Richard Nixon announced the roll back of the US program on development of biological weapons, "biological weapons have unpredictable and unavoidable consequences. They can cause serious epidemics and affect the health of future generations. Hence, I have decided that the US will renounce the use of lethal biological agents and all other methods of biological warfare". This declaration is the basis of the 1972 Convention.

Russia

In 1946, the Soviets created a new centre at Sverdlovsk, for studies on microbes of interest such as those causing plague, anthrax, tularemia, typhus, botulism and many others. A little later, in 1952, an open air research centre was opened on the island of "Renewal", Vozrojdenie on the Aral Sea, which is currently facing an ecological disaster caused by the gradual drying of sea due to diversion of two rivers that fed into it. Last, but not the least, in 1973, the Biopreparat was created with the aim of "developing genetically modified pathogenic agents, resistant to antibiotics and vaccines...". This activity involved 60,000 people, including 6,000 top level scientists.

In 1979, sixty-eight people died in an incident in Sverdlovsk (now Yekaterinburg), which strained the relations between Moscow and Washington during the 1980s. The Reagan administration accused the

Soviet authorities of using mycotoxins ("yellow rain") against their enemies in the south-east Asia which was in direct contradiction of the Convention Biological Weapons Convention of 1972 that had been signed between the United States and the Soviet Union.

The Russians responded angrily to these allegations invoking consumption of spoiled meat as the cause of the deaths. An article in the Tass Agency on 24th March 1980 tried to demonstrate that the infection was due to the anthrax bacillus which was naturally present in this endemic region. The CIA sought the aid of Mathew Meselson, a Harvard specialist in biological weapons to investigate the incident. He concluded that the symptoms did not correspond to intestinal anthrax in the first place and ruled out the hypothesis of ingestion of spoilt meat. It was only after the fall of the Soviet Union in 1991 and the accession to power of Boris Yeltsin, that the truth was finally made public. Yeltsin was the Chief of the Communist Party in the region of Sverdlovsk at the

A victim of the Sverdlovsk incident.

time of the incident. He was convinced that the KGB and the military had kept from him the real nature of the incident. In February 1992, at the summit meet between the two countries, President George Bush announced that the American accusations were well founded. In June 1992 Meselson led a research team to Sverdlovsk (followed by another visit in 1993) to complete the investigation. He was allowed to examine slides of lung autopsies performed on the victims which showed unambiguously that anthrax had been inhaled and not ingested.

Meselson published the results of his investigations in the journal "Science" in November 1994. In a military building close to the city, a small quantity of anthrax spores had been released in the atmosphere, after removing a filter. The exact quantity of the spores could not be determined, but could have been at least 1 gram according to Meselson, or several kilos according to the US Defense Intelligence agency. According to the Soviet authorities, a small quantity of spores had caused the death of 68 people and eliminated a herd of goats grazing in a meadow located 50 km away. The US sources claimed that the number of affected was more than 1000.

South Africa

In 1981, six years after having ratified the Biological and Toxin Weapons Convention (BTWC), the South African government headed by Pieter Willem Botha undertook a program of developing offensive weapons,

both chemical and biological, under the code name, "Operation Coast". Various biological agents were produced, including the spores of anthrax, bacilli of cholera and typhoid, and the botulinum toxin. These were used with success on several occasions.

The program was finally ended in the 1990s when the African National Congress succeeded in coming to power in the Government that was held by the minority white population. The project leader, Colonel Wouter Basson, a cardiologist and officer in the South African army was implicated in more than 60 major allegations. Basson had a fertile imagination and had proposed plans for the use of Ecstasy or LSD, syringes hidden in screwdrivers, chocolate spiked with anthrax spores or cyanide, beer mixed with botulinum toxin, etc. He was also interested in designing weapons based on the genetic differences in skin pigmentation, to specifically eliminate the blacks. The trial lasted 4 years and ended with the acquittal of Basson of all charges.

Colonel Wouter Basson, the "doctor of death", (AP/Sipa Press).

Iraq

Iraq started its program on biological weapons in the mid-1980s. When the Gulf War broke out in 1991, Iraqi scientists had already developed weapons containing anthrax spores, botulinum toxin, aflatoxin (a mycotoxin produced by a microscopic fungus called *Aspergillus*) and had several other similar projects underway. The Iraqi laboratories had produced 8,000 litres of a suspension anthrax spores at a concentration of one million spores per milliliter and 6,000 litres had been already packaged in the R400 bombs, each containing 85 litres of the suspension. They had also produced 20,000 litres of botulinum toxin, of which 12,000 litres was found after the end of the War along with 200 litres of aflatoxin. In all two hundred R400 bombs, including 50 loaded with anthrax spores, and 25 Scud ballistic missiles of 300 km range, loaded with biological agents, had been deployed by the Iraqi army in 1991 war. Fortunately, the Americans managed to ground the Iraqi aviation and thus prevented the transmission of the biological weapons.

Saddam Hussein (1937 – 2006), President of Iraq in 1979 – 2003.

From the Dark Ages to Modern Times

8 Attempts at disarmament

There have been numerous attempts at disarmament during the last century. The first took place soon after WWI with the signing of the Geneva protocol. Later, after taking note of the excesses committed during the WWII, there was a stronger political will to boost disarmament, which culminated in the 1972 convention.

The Geneva Protocol, 17th June 1925

This protocol was proposed after WWI. It reinforced the earlier declarations of 1899 and 1907 calling for prohibition of use of poisons during times of war. It also prohibited use of asphyxiating gases, toxins and all other similar biological means. By signing this protocol the participating countries agreed to recognize and extend the ban to biological weapons. The term "bacteriological" that covered viruses, rickettsia, bacteria and fungi, was used synonymously with "biological" and no particular species was mentioned. It is the first multilateral accord that extended the chemical ban to biological agents. The signatory countries were bound by this declaration and promised to try and convince others to sign the Protocol. All newcomers were to notify the French government, which would in turn notify the other signatories and members.

During the negotiations for the Protocol, most of the countries talked about banning first-use of the biological weapons. They wanted nevertheless, to keep the option of using biological and chemical weapons in retaliation of an attack of this nature. In order to deal with these eventualities, the signatory countries therefore needed to develop, manufacture and stock these arms, which was not bound by the Geneva convention. Curiously, this led to a proliferation of these weapons in countries that had a research program. In May 2003, 133 States had ratified the program, while some were actively engaged in developing biological weapons at the time. Also, the issue of monitoring compliance with the protocol had not been discussed.

The Biological and Toxin Weapons Convention (BTWC): 1972

"Recognizing the important significance of the Protocol (...) signed at Geneva on the 17th June 1925, the State Parties are determined to act with the view to achieving progress towards general and complete disarmament, under strict and effective international control," (excerpt of the official text of the convention signed on the 10 April 1972). All the countries present at the Convention pledged never to develop, manufacture, stock or acquire in one way or another, any bacteriological or biological agent or their toxins, regardless of their origin, type or quantity. The first article of the Convention however stressed that this did not include agents "intended for prophylaxis, protection or other peaceful purposes", but condemns weapons, vectors and all kinds of equipment that would facilitate use of these agents.

All the signatory countries had to monitor and ensure the implementation of this in their own territory and had the right to lodge a complaint with the Security Council of the United Nations against any party that violated this agreement. Similarly, they would not have any alliance with or help in any way whatsoever, any State or organization, that violates this Convention.

By this Convention, the States undertook to consult and cooperate among themselves to resolve all problems that may arise. They also promised to share equipments, material, scientific and technical information concerned with the prevention of diseases and other peaceful ends.

The Convention met in Geneva five years after its signature in the order to take stock of its implementation and to verify that the objectives had been achieved. It stressed that it welcomed all countries and that it would stay in effect for an indefinite period. However all the signatory states had the option of withdrawing from the agreement if they thought necessary. At the time of its ratification on the April 10, 1972, 103 countries had signed this Convention, and in 1997, there were 146. This indicates the international awareness concerning this risk and the willingness to take steps to cope with it.

However, this convention was restricted to the development and use of biological weapons. It "overlooked" the problem of monitoring and interpreting defense research. As such the Protocol did not discuss the small quantities of the stock cultures aimed at conducting research on the means of defense in case of an attack. Moreover, there was provision for retribution in case of violation of the Convention. Lastly, this convention applies essentially to countries, while there is a greater risk from terrorist groups and organizations, which cannot be brought to a negotiating table.

9 The lure of bioterrorism

Rajneesh sect and salmonellas

Bhagwan Shree Rajneesh (1931 – 1990), a firm believer in Hinduism, created the "Rajneesh International Foundation" or the Rajneesh sect and was one of the most controversial gurus of his time. The sect had to leave their headquarters in Pune (earlier, Poona), in western India, after various irregularities and relocate to the US.

In 1981, at a cost of 5.75 million dollars the sect acquired a 32,000 hectare, isolated ranch, in the rural county of Wasco in Oregon, about two hours by car from the little town of The Dalles. The Rajneesh community expanded pretty rapidly and attracted nearly 4,000 followers. Soon the Rajneeshees took

Bhagwan Shree Rajneesh.

control of Antelope, a little town with 75 living souls, and changed its name to Rajneeshpuram. The next step was to broaden their zone of activity to include the entire county by means of the local elections in 1984. They had plans to infect a large proportion of The Dalles population or a majority of the 20,000 inhabitants to prevent them from voting. The members of the sect contaminated the salads and other raw foods in 10 restaurants in The Dalles with a culture of *Salmonella*. The idea was that in case this plot succeeds, then the same technique can be used to introduce the bacteria in the drinking water supplies of the town on election day. Finally, they managed to infect more than 700 people. The results of the investigations placed the responsibility of this first ever bioterrorist attack on American soil, on the Rajneeshees. In 1985, Rajneesh was deported from the US on the grounds of illegal immigration. Twenty-one countries refused him hospitality and he was forced to return to Pune. The sect was reestablished and again became popular gathering 15,000 members.

Aum Shinrikyo sect

Shoko Asahara.

This apocalyptic sect was led by Shoko Asahara, a practically blind guru. It became notorious for executing the sarin nerve gas attack in Tokyo subways, killing 12 and injuring 5000 others, on 20 March 1995.

The members of the sect were also accused of having perpetrated other acts of bioterrorism – six using botulinum toxin and four with anthrax spores. In 1990, they dispersed botulinum toxin near the Japanese parliament and a few months later anthrax spores were sprayed as aerosols from the top of a building in Tokyo. Fortunately, none of these attacks had any visible consequences, possibly because the culture of *Bacillus anthracis* used was not sufficiently virulent or the system of dispersion may not have been adequate.

Anthrax letter bombs of October 2001

Letters containing anthrax spores, sent to Senator Daschle and Tom Brokaw in October 2001.

4th October 2001, Palm Beach County Health Department, Florida: a death is announced. Robert Stevens, 63 years old, photographer of the daily "Sun" dies of anthrax poisoning. Apparently he had been exposed to the anthrax spores just before going on a vacation on the 26th September. He began to feel sick by the 30th and was admitted to the JFK Medical Centre in Palm Beach on the 2nd October. Four hours later, he was comatose. Robert Stevens was the first victim of the anthrax letter bombs that followed the events of 9th September 2001 (now simply known as "9/11"). Over a period of few days, four other people in New York and Washington died from similar mail bombs, and seventeen others were contaminated. In all, it cost about one million dollars to decontaminate affected buildings like the senate and the postal services and cleanse them of the anthrax bacillus.

In September – October 2001, about 5 – 6 letters containing the spores of *Bacillus anthracis* were posted from Trenton, New Jersey. These letters contained identical messages trying to prove that the authors were Islamic terrorists. They said, "9/11: this is next. Take penacilin (*sic*) now. Death to America". Since these letters went through automatic machines, they infected several thousands of postal employees and once opened hundreds of others. It is almost certain that many cases were prevented by a prophylactic antibiotic treatment given to people who had been exposed. The authors of these letters is still unknown, but a strong suspicion was placed on Steven J. Hatfill, an ex-employee of the US biodefense program. He was a biological weapons expert who had spent several years in Rhodesia (today Zimbabwe) at the time of an anthrax outbreak. It turned out that Hatfill had been wrongly accused and the real culprit was an American microbiologist, Bruce E. Ivins. Latter committed suicide on 29th July 2008 during his court hearing. As for Hatfill, whose professional and personal life had been completely destroyed since 2001, he received a compensation of 4.6 million dollars.

Steven J. Hatfill (Source AP/Sipa Press).

Glossary

Major pathogenic agents

Microbe

Microbes or microorganisms, are living organisms that are invisible to the naked eye and can be seen only through the lens of a microscope. Their size is in the order of a micrometer or one millionth of a millimeter. In general, one can distinguish bacteria, viruses (100 times smaller than the bacteria), yeast, fungi, certain parasites and even some of the newly found infectious agents like the prions.

It was Antonie van Leeuwenhoek, a Dutchman, who in 1673 first described "animalcules" that he observed by using a magnifying glass on the mesh of sheets that he used to sell in his drapery business.

Microbes can be used as ferments to produce yoghurts, cheese, bread, choucroute, beer, etc. They can also be responsible for infectious diseases:

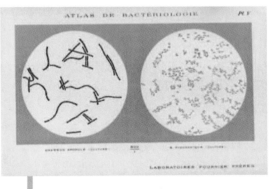

Microbes seen under a microscope.

- Bacterial: whooping cough, scarlet fever, diphtheria, cerebrospinal meningitis

- Viral: smallpox, rubella, herpes

- Parasitic: toxoplasmosis, pinworms

- Prion: mad cow disease, Creutzfeldt-Jakob disease

Bacteria

A bacterium is a unicellular micro-organism composed of a membrane and a cell wall. Its size is variable but the elongated bacteria (bacilli) measure around 3 – 4 micrometer while the round bacteria (cocci) have a diameter of about 1 micrometer.

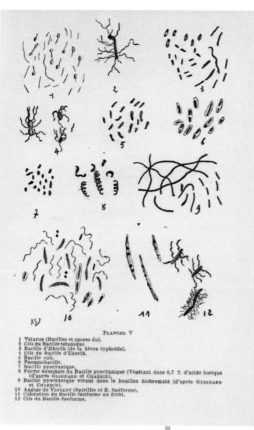

PLANCHE V

1 Tétanos (Bacilles et spores du).
2 Cils du Bacille tétanique.
3 Bacille d'Eberth (de la fièvre typhoïde).
4 Cils du Bacille d'Eberth.
5 Bacille coli.
6 Pneumobacille.
7 Bacille pyocyanique.
8 Forme anormale du Bacille pyocyanique (Végétant dans 0,7 % d'acide borique (d'après GUIGNARD et CHARRIN).
9 Bacille pyocyanique vivant dans le bouillon bichromaté (d'après GUIGNARD et CHARRIN).
10 Angine de VINCENT (Spirilles et B. fusiforme).
11 Coloration du Bacille fusiforme au Ziehl.
12 Cils du Bacille fusiforme.

Bacteria of different shapes.

The cell is bound by the membrane. Inside the cell are: the nucleic acid, the carrier of genetic information and different components enabling it to produce energy from various substrates. Outside the membrane, a solid cell wall gives it its shape and protects it from fluctuations in the external medium conditions, most importantly, variations in the osmotic pressure.

Based on the different staining properties, two major types of cell walls can be distinguished. The Danish microbiologist, Hans Christian Joachim Gram, developed a staining procedure in 1884 that allows to differentiate between the cell wall types. Some bacteria, called "Gram positive" are stained violet in the Gram staining procedure, while those called "Gram negative" stain pink. Some bacteria do not take the stain at all (spirales, causing syphilis, *Treponema pallidum*), while others are highly resistant to destaining even under harsh conditions such as by treatment with acids or alcohols (for example, bacteria causing tuberculosis, *Mycobacterium tuberculosis*).

Some bacteria transform into a resistant form called spore, that allows them to resist treatment with acids, alkalis, antiseptics and disinfectants, as well as changes in temperature or pressure. Thus, even 4,000 year old spores found in the Egyptian mummies can revive and produce bacteria. The spores of *Bacillus anthracis* are a good candidate for a biological weapon. Some pathogenic bacteria produce a capsule around themselves and are thus protected from attacks from the host phagocytic cells that are supposed to eliminate them. Certain types of bacteria possess cilia and flagella on their surface, which allow them to move actively in their environment.

Viruses

A virus is a biological entity measuring on an average, about 100 nanometers (1 nanometer = 1 thousandth of a thousandth of a millimeter). It is composed of an outer, proteinaceous capsid, which protects its nucleic acid, the carrier of genetic information.

The virus is so tiny that it cannot carry the machinery required for its own replication. Hence, it borrows the machinery of the host cell which it invades. In most cases, it destroys the host cell when it escapes from it, thereby causing disease symptoms. Viral diseases include influenza, chicken pox, small pox, measles and yellow fever.

Some viruses have an envelope (enveloped viruses) which is assembled from the components it takes from the cell membrane of the host cell within which it multiplied. These viruses are more vulnerable (eg., viruses causing herpes, influenza) compared to others that do not have this envelope, called "naked viruses". Latter are much more resistant (eg., virus causing warts, hepatitis A).

The human immunodeficiency virus, (HIV), that causes the Acquired Immunodeficiency Syndrome (AIDS) multiplies inside cells called T4 lymphocytes. These cells coordinate our immune defense system. If these cells are missing, the body's defense is compromised and the individual becomes extremely vulnerable to infectious diseases. The patient is prey to even those infectious agents that are normally not aggressive and death may occur due to what are called opportunistic infectious agents.

Parasites

A parasite is a living organism that lives at the expense of the host at some point in its developmental cycle. The "definitive host" is the one in which the sexual development of the adults takes place. The asexual phases of the cycle, or the larval stages, can occur in the external environment or in other hosts, called "intermediary hosts" (mollusks, mosquitoes, fleas, flies). There are several diseases caused by parasites. Some of the diseases are better tolerated, for example, a few ascaris parasites in the intestine, while others are dangerous or even fatal, for example, malaria that causes 4 – 5 million deaths every year worldwide.

Yeast and fungi

Some of the diseases are caused by microscopic fungi. These are plants that do not have roots, stem nor leaves. They give out branching filaments. The simplest fungus is the yeast that consists of a globular single cell. When some fungi colonize the human body, they can cause diseases, collectively called mycoses. The best known examples is that of thrush, caused by the yeast, *Candida albicans*. This species opportunistically benefits from a lowering of host immune defenses to colonize the mucus coverings of the mouth or vagina, which at times can be extremely painful. Another fungus called *Aspergillus niger* can cause

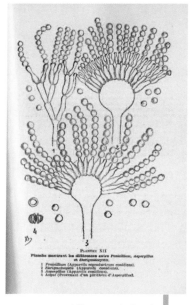

Microscopic fungi.

lung infections in severely immunocompromised individuals. Some of these fungi produce and secrete dangerous toxins called mycotoxins, for example, aflatoxin produced by fungi of the genus *Aspergillus*.

Major infections

Anthrax

Anthrax is caused by a Gram positive bacterium, *Bacillus anthracis*. It is an animal disease, which is not contagious and rarely affects man. Humans can be infected due to contact with diseased animals resulting in cutaneous anthrax (by skin contact), gastrointestinal anthrax (by ingestion of contaminated food) or pulmonary anthrax (by inhalation). The disease mostly affects farm animals such as sheep, cattle, goats and horses. The bacteria are capable of transforming into the spores, which are highly resistant to environmental stresses. The spores can survive in soil for a very long time after the diseased animal is dead and buried. The spores are brought to the surface by the activity of earthworms and can thus contaminate other animals that graze in the same field. This explains the term "cursed fields" applied to fields in which healthy animals routinely die of anthrax after grazing there.

Cutaneous anthrax: This occurs when a wound or a scratch gets infected with the anthrax bacilli from contact with a diseased animal's skin, hair or carcass. This form of anthrax is seen as papules that progressively develop into vesicles within 24 – 48 hours, turning into painful ulcers and sores that appear black. It derived its name of "carbuncles" meaning charcoal ("charbon" in French) from the characteristic black ulcers. Serious complications or death are rare (around 5% of the cases). Cutaneous anthrax caused by *Bacillus anthracis* should not be confused with another skin disease caused by *Staphylococcus aureus*.

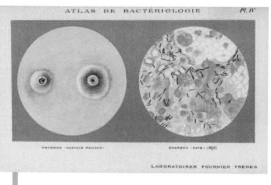

Malignant pustule (left) and microscopic observation of the anthrax bacillus (right).

Gastrointestinal anthrax: This occurs from eating contaminated meat that is insufficiently cooked. Symptoms are gastroenteritis, fever, nausea, vomiting, difficulty in swallowing, abdominal pain, and severe

From the Dark Ages to Modern Times

diarrhea with blood in the stools. The disease progresses rapidly and can end in death if not treated. This form is now rare since the introduction of proper vaccination of the livestock.

Pulmonary anthrax: This form occurs from inhalation of the spores of anthrax bacillus. The incubation period before which the symptoms appear may vary between a day to two months. This is a serious disease, often lethal, that starts with flu-like symptoms of fever, headache, malaise, muscular pain, cough and breathing problems. Within a week, the disease progresses into respiratory distress, cardiac failure and coma. Once inside the host lungs, the spores germinate and the bacteria release a deadly toxin that destroys the lung tissue and may even reach the brain.

Spores of *Bacillus anthracis* have been used in aerosols in several bioterrorist attacks. The most important features concerning this choice is the highly mortality caused by this form of anthrax and the fact that the highly resistant spores can survive for long periods of time.

All three forms of anthrax are treated with antibiotics. In case of the gastrointestinal form, treatment has to start very soon and for the pulmonary form within 48 hours of inhalation, that is to say, before the toxin is released.

An effective animal vaccine is available. The human vaccine however, is not satisfactory and has considerable side effects. Hence it is not used routinely given the low risk of an airborne contamination.

Botulism

Botulism is rather rare in France, occurring at the rate of about 20-odd cases per year. It is an intoxication, that is to say it comes from ingestion of a toxin released by *Clostridium botulinum* in contaminated food. It is not a contagious disease. *Clostridium botulinum* is a Gram positive bacterium that is a strict anaerobe, meaning, it cannot survive in the presence of oxygen. It lives in the soil and is capable of transforming into the resistant form of a spore. It synthesizes a neurotoxin that inhibits the communication between nerves and muscles and thus induces paralysis. Soon after ingestion of food contaminated with botulinum toxin, the first signs to appear are problems in vision followed by a general muscle fatigue. In serious cases, respiratory disorders can cause death. Foods likely to contain this toxin are canned preserves such as asparagus, French beans, beetroot, corn or ham. Canned meat (ham for example) may contain the toxin if the animal was not on an empty stomach at the time of its killing. After a meal, the bacteria residing in the animal's intestine can enter

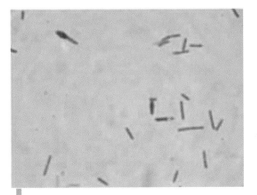
Clostridium botulinum: Gram staining.

the bloodstream and reach the muscles where they multiply and synthesize the toxin under anaerobic conditions after the animal's death. Recently, a batch of "foie gras", a French specialty, typically made from liver of a fattened goose or duck, was withdrawn from the market after traces of botulinum toxin were detected.

The toxin can also enter the blood directly through an open wound and give the exact same symptoms. This form of infection is dangerous in cases of microbiological warfare, because it can bring a quick death.

Currently, the toxin is used under strict medical supervision to induce certain specific muscles to relax and aid in erasing wrinkles at the site of injection: popularly known as botox.

Cholera

The causative agent, *Vibrio cholerea*, is a Gram negative bacterium. Man is the principle host and the disease is spread through infected food and water.

Worldwide distribution of cholera.

Today it is prevalent mainly in Africa. War and movements of refugees favor epidemics, when hygienic conditions are not satisfactory. In July 1994, about 24,000 Rwandans died in refugee camps. Similarly, between January and May 1998, 13,500 cases of cholera and close to 800 dead were registered by the WHO in the Democratic Republic of Congo. In the same period, 20,000 cholera cases and 1,000 dead were reported in Uganda.

The cholera bacteria are ingested through food or water and enter the gut, where they multiply and release a toxin. It is the toxin that induces the disease symptoms, particularly a huge loss of water. The infected patient becomes dehydrated, at times losing 15 liters of water in a day from severe diarrhea. There is no fever, but intense abdominal pain and severe diarrhea and vomiting. Patients who are already weak can die in 1 – 3 days if they do not receive treatment in the form of rehydration. Antibiotics are usually ineffective. The feces of the patient carry the bacteria to the environment, where they spread through the fecal-oral route to contaminate other people.

Dengue fever

Dengue fever is an infectious disease caused by a virus and it continues to gain new ground even today. It is found in most cities in the tropical regions of the world, such as islands in the Pacific, Asia, Africa and the Americas. In all, two-fifths of the world population is concerned with

From the Dark Ages to Modern Times

dengue fever and 50 million cases are registered annually by the WHO.

The disease is transmitted from man to man mainly by the principle vector, the mosquito, *Aedes egypti*. This mosquito may bite the same person several times thus transmitting the virus. The disease begins with

Worldwide distribution of dengue fever.

a high fever, chills and shivering, headache, muscular pain and a general malaise. The sickness may simply stop here, without leaving any mark. Or, after a more or less brief remission, the condition may deteriorate and other symptoms may be seen: hemorrhagic patches on the skin, and bleeding from the nose and gums, signaling coagulation disorders. In severely hemorrhagic cases, resuscitation may be required. There is no vaccine against dengue fever; protection from mosquito bite is the essential preventive measure.

Dysentery

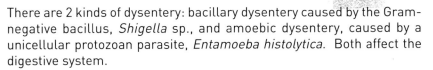

There are 2 kinds of dysentery: bacillary dysentery caused by the Gram-negative bacillus, *Shigella* sp., and amoebic dysentery, caused by a unicellular protozoan parasite, *Entamoeba histolytica*. Both affect the digestive system.

Bacillary dysentery or shigellosis

Shigellosis is currently prevalent in tropical countries where is flares up during hottest and wettest seasons of the year. Natural disasters, wars, people movements, are all favorable to outbreaks of the disease. An estimated 165 million cases are recorded annually worldwide, of which more than a million die, especially children less than 5 years.

The bacteria reside in the human digestive tract. Contamination is therefore via the fecal-oral route, either by direct contact (unwashed hands) or indirect (fecal contamination of food or water; or fruits washed with contaminated water). The ingested bacteria invade host intestinal cells and destroy them. Severe and frequent diarrhea, tinted with blood results. Other symptoms include fever, uneasiness, violent abdominal cramps and vomiting.

If a diarrhea, principally consisting of water, is self-limiting, then it is the benign form of shigellosis. In severe cases serious complications may arise such as intestinal bleeding or perforation of the colon; or general complications such as dehydration, septicemia (presence of bacteria in the blood), neurological complications, acute renal failure and coagulation disorders. These complications can lead to death in children. Patient should be rehydrated and given antibiotics.

Amoebic dysentery

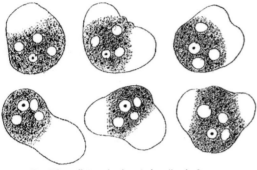

FIG. 259. — *Entamœba dysenteriae*, d'après Jurgens.

Entamoeba histolytica: chemical, microscopic and parasitological diagnosis.

The causative agent of this disease is the unicellular parasite, *Entamoeba histolytica*, that replicates in the human intestine and is eliminated in the feces, and thus transmitted by the fecal-oral route.

The disease is seen all over the world, with the incidence rates being very high in regions where the hygienic conditions are poor. The highest number of cases is seen in tropical regions, where it is difficult to maintain a high standard of hygiene. In contrast to shigellosis, amoebic dysentery is less frequent in children of less than 5 years. Every year there are about 50 million infections causing 70,000 deaths, most often due to complications, especially liver abscess.

Like in the case of bacillary dysentery, this disease also spreads via the fecal-oral route, either by direct contact (unwashed hands) or indirect (fecal contamination of food or water; fruits washed with contaminated water). The symptoms include frequent diarrhea, blood in the stools, vomiting, abdominal cramps and fever. In some case, the parasites can invade the circulatory system and infect the liver, lungs and brain. Death may occur, principally from liver abscess, in patients who are already weak.

Ergot of rye

Ergotism or "St. Anthony's fire" or " burning sickness" is due to ingestion of rye infested by the ergot fungus, *Claviceps purpurea*, which produces toxic substances called alkaloids. In the past, there have been outbreaks of severe poisoning called "ergotism" due to infestation of crops of rye by this fungus. It was also called "burning sickness" because of the burning sensation felt by the patients. The symptoms of the disease are characterized by hallucinations (like those given by LSD, which is a synthetic derivative of the fungal alkaloid), a loss of sensation in the extremities, especially the fingers, leading to gangrene and even death. The alkaloid causes arteriolar vasoconstriction or contraction of small arteries, preventing normal blood flow.

In the Middle Ages, healing from ergotism was attributed to Saint Anthony, who became the patron saint of victims of this sickness. Often the victims went on pilgrimage to the

Ergot of rye.

relics of Saint-Antoine-l'Abbaye (Isère). While travelling, they would not eat any more of the infected rye bread and thus returned cured.

Fowl cholera

This disease is caused by a Gram negative bacterium, *Pasteurella multocida*. It is one of the most important species in veterinary medicine. The bacteria are found in the mucous membranes of respiratory and digestive tracts of a large number of animals and are opportunistically pathogenic if the animal is under stress. These bacteria can cause cholera in chicken, atrophic rhinitis in pigs and rabbits and a variety of other diseases in different animals such as cattle. The bacteria usually enter through the respiratory route and then get into the bloodstream, which is called septicemia. The birds' feathers get ruffled, it has fever, difficulty in eating, and a diarrhea that is initially white and then green before death. Sometimes, the animal is found dead in the cage if the breeder does not notice the clinical signs. Epidemics are rare, but mortality is very high, especially in turkeys.

The first vaccine was developed by Pasteur through the attenuation of virulence of the bacterial strain. It was also Pasteur's idea to spray a culture of this bacterial strain in rabbit feed in order to control their population (see the section: Pasteur "precursor of biological weapons"). These bacteria were thus the source of an epidemic in these animals, called "epizootic" and represent one of the early microbiological weapons.

Chicken,
a host for Pasteurella mulocida.

This disease should not be confused with the human cholera which is caused by *Vibrio cholerae*.

Gas gangrene

This is a serious infection caused by the bacterium, *Clostridium perfringens*. These bacteria are strictly anaerobic, i.e., survive only in the absence of oxygen, and produce a large number of toxins. These are found in the soil and the digestive tract of man and animals. An open wound or a fracture from a road injury for example, or a post-operative wound, can get infected by these bacteria. In 2 – 3 days, a gaseous abscess is formed that affects the muscle surrounding the site of infection. The bacteria produce a lot of gas making the affected area foul smelling and palpable. Muscle necrosis ensues. High fever and death can occur rapidly if left untreated. In the past, when there were no antibiotics, the infected region had to be amputated. The disease was common during wars. In women, genital infections may occur from use of contaminated material during abortions. Death occurs when the bacteria get into the bloodstream, the phase called "septicemia".

Today, gas gangrene can be cured by an effective antibiotic therapy.

Glanders

Glanders is a contagious disease found mainly among equines. Humans can contract it by direct contact with sick animals through skin weakened by microabrasions in the mucous membranes of the mouth or by nasal inhalation. It is caused by a Gram negative bacterium, *Burkholderia mallei.*

Face of a patient of glanders (a student from Alfort, who died in 1836), Museum of Maison Alfort.

In equines, pulmonary glanders is characterized by a runny nose (nasal discharge) with pus and abscess in the bronchi of the lungs. Skin glanders seen as ulcers that do not heal, and contain pus. Both types of glanders are highly contagious in livestock. In humans, the infection is mostly found around a wound. After an incubation period of ten days to a month, there is local suppuration accompanied by swelling of the glands. The general condition worsens with fever, fatigue and emaciation. If untreated, the infection can spread and form of abscesses in other organs including spleen, liver and lungs. The disease progression is slow, occurring over a period of three weeks to three months, but often lethal. The acute form has an estimated mortality rate of 95%.

Due to the high mortality rate and the small quantity of germs required to cause an infection, *Burkholderia mallei* and the closely related *Burkholderia pseudomallei*, the agent of meliodosis, are considered as a potential biological weapons.

Influenza

Influenza is caused by an enveloped virus. The surface of the virus has two types of structures (antigens): one called "H" and the other "N". The flu viruses are distinguished on the basis of combinations of H and N surface antigens. The H2N2 virus was responsible for the Asian flu epidemics of 1889 and 1957; H1N1 for the Spanish flu of 1918 – 20 that caused about 40 million deaths; and H3N2 for the Hong Kong epidemic of 1968. The disease occurs in birds and animals as well. Currently, the virus H5N1, responsible for the "bird flu" has not infected humans yet, but its high virulence warrants a close surveillance of its progression.

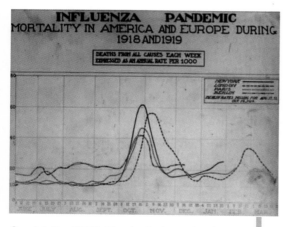

Spanish flu of 1918. The graph shows that the peak of the infection occurred in the winter of 1918.

The disease is transmitted by infectious droplets released by the coughing or sneezing of the patient.

Disease symptoms include a sudden onset of a fever, chills, headaches, pains in the joints and muscles. There may be irritation in the eyes, throat and chest. The fever is general limited to 3 – 5 days, but fatigue and weakness may persist for several weeks.

In subjects who are fragile (the old or immunocompromized), respiratory failure may occur, leading to death (about 3,000 deaths per year in France). Treatment with antiviral drugs is not currently practised in France, however people aged more than 65 years and frail people are regularly vaccinated. As the virus tends to change its surface structures (from mutations), it is not exactly the same virus from one year to another. Consequently, the vaccine also has to be changed annually. If infected, the doctor may prescribe antibiotics (these do not work against viruses) to prevent secondary infections by bacteria.

Leprosy

Leprosy is an infectious disease caused by *Mycobacterium leprae*. These bacteria are highly resistant to acids and alcohols and cannot be classified by the Gram staining method. These are called BAAR for Bacilli acid alcohol resistant. It is estimated that around 15 million people are affected by leprosy worldwide, and 2 million suffer a disability due to it. The countries most affected are located in the tropics: Black Africa, China, East Asia, India, Madagascar, Portugal, Spain, Caribbean and South America. Poor hygiene and overcrowding, promote the spread of the disease.

The infection progresses very slowly. The first symptoms appear only 2-8 years, sometimes 20 years, after infection. The leprosy bacilli, also known as Hansen's bacilli, are transmitted through droplets from the mouth or nose of patients. The disease mainly affects the skin and nerves. Without proper treatment, the lesions extending to limbs and eyes become permanent. There are different types of leprosy. Tuberculoid leprosy is non-contagious and characterized by the appearance of large discolored spots on the skin (less than 5). There is a loss of sensation, skin feels anesthetized which leads to disability and

Leprosy patient with multiple disabilities.

Leper ringing a bell
(Latin manuscript of 14th century).

maiming. Muscle disorders are accompanied by nervous disorders. Lepromatous leprosy is contagious, causes lesions on the skin (more 5), mucous membranes and affects several organs, including the face, liver, spleen, nerves and bones. A purulent and bloody rhinitis is characteristic of this type of leprosy. A spontaneous progression is driven by successive painful and febrile spells leading to death in 10-20 years.

Leprosy is curable with a mixture of antibiotics (polychemotherapy), the only drawback being that the treatment period is very long(2 years).

Malaria

This is the most prevalent parasitic disease in the world with over 500 million patients and several million deaths per year (mostly children). A majority of the cases (85%) occur in sub-Saharan Africa. We also find it in rural Cambodia, Indonesia, Laos, Malaysia, Philippines, Thailand, Vietnam and China in Yunnan and in Hainan, Haiti and in rural areas in Bolivia, Colombia, Ecuador, Peru and Venezuela, and throughout the Amazon region. In Central America, the risk is relatively low.

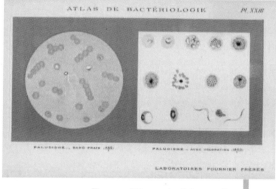

Forms of the malarial parasite, from "Atlas of bacteriology".

Malaria, meaning "bad air", is caused by a parasite that is transmitted to humans by a female mosquito of the species *Anopheles*. The mosquito injects the infectious form of the parasite through its bite. With the bloodstream the parasites reach the liver and multiply in the liver cells, which constitutes the first phase of the disease. The liver cells burst releasing the parasites which then attack erythrocytes or red blood cells (RBCs) and go through a second phase of multiplication to produce the sexual forms of the parasite. When the infected RBCs burst, other RBCs get infected and this erythrocytic cycle is repeated several times, causing the characteristic fever with every round of multiplication. This phase is thus characterized by a fever that follows a three-day cycle, i.e., fever appearing on days 1, 3, 5, etc or a four-day cycle, i.e., fever appearing on days 1, 4, 7, etc. A malarial attack or clinical malaria can consist of a dozen successive cycles of fever or may be repeated over several years. The fever of around 40°C and chills, are followed by a drop in body temperature and abundant sweating and feeling of cold.

The blood borne sexual forms of the parasite are picked up again by a mosquito during a blood meal on a malaria patient. The parasite undergoes developmental changes in the mosquito to produce infectious form, thus completing its life cycle.

There are four different types of parasites, the most dangerous being *Plasmodium falciparum* that causes the three day fever cycle and sometimes death. It can cause what is called 'cerebral malaria" which is lethal in 20% of the cases after going through a coma.

The treatment is based on chloroquine or primarquine. Sometimes treatment with quinine may prove dangerous as it can cause RBCs to burst inside the vessels.

Protection against malaria consists of eliminating mosquitoes, for example with the use of insecticides, use of mosquito-nets and preventive anti-malarial treatment with chloroquine in endemic regions. Given the complexity of the malarial cycle, there is currently no anti-malarial vaccine available.

Measles

This is an infectious viral disease. It occurs mostly during winter and spring in temperate regions and is transmitted by droplets of saliva from infected individuals. The eruptive fever affects mostly children worldwide, but unlike in developed countries, where its consequences are not serious, in developing countries, it causes high mortality (345,000 deaths in 2005). After an incubation period of about 10 days, symptoms of high fever are seen, accompanied by swollen eyes, conjunctivitis, runny nose, cough and abdominal pain. A doctor can observe numerous white spots on the insides of the cheeks called the Koplik spots. A few days later these disappear, to be replaced by the characteristic rash on the skin: small red blotches that fuse to firm larger blotches. The affected organs include

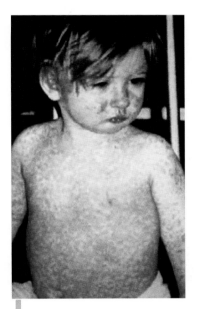
A child infected with measles virus.

ears, neck, thorax, abdomen and the limbs. In the eruptive phase of the disease, the body temperature stays high and the nasal and eye symptoms appear. The child appears tired and its immune system is weakened. If all goes well, there are no complications, otherwise, lungs and brain can get affected and secondary bacterial infections may occur. These complications make measles a serious infection.

An effective vaccine is available that has resulted in the reduction by 75% the number of children dying in Africa between 2000 – 2005. In France and in other countries, the triple vaccine MMR (measles, mumps, rubella) is recommended at the age of 1 year with a booster dose a year later.

Meningitis

Meninges are the protective layers surrounding the brain that prevent it from being in direct contact with bones of the skull. When these are colonized by infectious agents, it is called "meningitis". Depending on which agent is causing it, meningitis can take different forms. In general, viral meningitis (80% of the cases) is benign while the bacterial is more severe (20%) and can cause death. The bacteria most frequently involved are the pneumococci, streptococci, staphylococci and *Listeria monocytogenes*. Currently, there are few incidences of tubercular meningitis and that caused by *Haemophilus influenzae* as a result of vaccination of young children.

The most dangerous meningitis is without doubt, the one caused by *Neisseria meningitidis* or the meningococci. These bacteria can trigger small outbreaks in nurseries and schools. There are four major types of bacteria, called A, B, C and the W135, which is prevalent in Mecca. The disease is spread through contact with secretions from the nose and throat of infected persons. The bacteria are sensitive to cold and cannot survive in the outside environment.

The symptoms of bacterial meningitis develop rapidly, within two days or may even cause death in just a few hours. Infants do not display the characteristic clinical symptoms of the disease. In children older than 2 years, adolescents and adults, there is a high fever, inability to eat or drink, vomiting, severe headache, stiff neck and hypersensitivity to light. A purple colored skin rash or *purpura fulminans* is a sign of severe meningococciemia or presence of the meningococci in the blood stream. If untreated, it leads to death in 20 – 30 % of the cases.

Diagnosis is based on analyses of the fluid that bathes the meninges, called the cerebrospinal fluid (CSF), samples of which can be drawn from lumbar vertebra for testing. The bacterial cells causing the meningitis are found in the CSF sample. Treatment consists of injection of antibiotics as soon as possible. Persons in contact with the patient are also given the antibiotic treatment as a preventive measure.

A vaccine is available against meningococci of the groups A, C and W135 but not against the group B. In France, only people in contact with sick children and those visiting endemic countries are vaccinated.

Plague

This disease is caused by a Gram negative bacterium, *Yersinia pestis*, named after Alexander Yersin who discovered the bacteria in 1894.

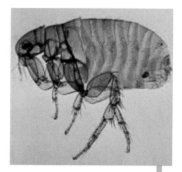

A flea observed under a magnifying lens.

The disease is disseminated by rodents: rats in Europe, squirrels in the USA. The bacteria are actually transmitted to humans by the fleas and mites from the rodents during a blood meal. While feeding on human blood, the fleas make a skin abrasion through which they release the contents of their gut, that may carry the plague bacillus. After a week of incubation period, there is high fever and the infected ganglions are swollen (adenopathy). Soon the bubos (enlarged lymphatic nodes) appear, usually around the groin and armpits. This is the

From the Dark Ages to Modern Times

bubonic plague. In half of the cases, patient dies within a week of the appearance of the bubos.

In some cases, the bacteria get into the bloodstream and invade all organs. This is the septicemia plague.

The third form of plague, which is the most serious is the pneumonic plague. In this case, the transmission of the bacteria is not by the bite of a flea but these are inhaled along with contaminated particles (sputum, sneeze) from an infected persons. A deadly pneumonia follows in a few hours.

Potato mildew

This disease of plants is caused by the fungus *Phytopthora infestans*. The bacteria proliferate under conditions of high humidity and temperatures around 17 – 20°C. The upper surface of the leaves show small discolored spots that turn brown and are surrounded by a yellow halo. On the underside of the leaves bordering the spots there is a characteristic white matty fungal growth. The number of spots increases and their spread causes dryness and a rapid destruction of the foliage. The potatoes have brown or bluish gray spots. A cross-section of an infected tuber shows marbled and rust colored areas extending from the surface to the center.

The great Famine of 1845 – 49 in Ireland was a consequence of a mildew that destroyed potato plantations.

In all cases, the control measures have to be preventive: use only healthy plants, destroy the ones contaminated and spray products to prevent the disease. There are several species of the fungus that cause mildew of different plants such as vines, sunflower, strawberries, peppers, cabbage, garlic, onions, spinach, tobacco etc.

Rabies

This is a viral disease that poses a serious problem in countries such as East Africa, Asia, Eastern Europe, China (50,000 cases/year), Thailand (300 cases/year) or South America (Mexico: 10 - 30 cases/year). Rabies still kills about 40,000 people each year worldwide. Since 1968, when it reappeared in France, rabies killed more than 45,000 animals, 35,000 foxes, 3,500 cattle, 1,600 cats and 1,000 dogs. The disease is extremely dangerous because once the symptoms are visible, it leads to death in all cases.

The fox, wolf, badger, deer, as well as domesticated animals like dogs, cats, cow can transmit a rabies infections to humans. In the case of carnivores, abnormally aggressive behavior may be observed. In this case, the animal tries to bite into anything in the vicinity of its head, and does not let go. The virus is present in all secretions from the infected animal, especially the saliva, and feces. Since the virus is extremely vulnerable in the external atmosphere, its entry into host relies on entry through broken skin, or mucous membranes of the mouth or eyes. It may also enter by inhalation of stale air in caves inhabited by infected bats.

Bat.

Rabies is most often transmitted by biting and much less frequently by licking. After infection, it invades the peripheral nervous system in humans. It then travels along the nerves to reach the central nervous system. Once inside the brain, it causes an encephalitis and the characteristic symptoms appear. It can also pass into the spinal column. Typically, there is lack of coordination of muscular movements (ataxia), anxiety, confusion, hallucinations and insomnia. There can be neck pain and convulsions of facial muscles. Advanced symptoms include production of a large amounts of saliva and tears and difficulty in swallowing. In humans, there is also hydrophobia in late stages and contact with water causes an unbearable burning sensation. Death follows inevitably within 2 – 10 days of the first appearance of symptoms.

Treatment consists of injection of the anti-rabies serum and post-exposure prophylaxis. A preventive human vaccine is available and mostly reserved for professionals like veterinarians and those visiting endemic regions. The animal vaccine is compulsory for domestic animals like dogs and cats. Since 2001, France is considered to be free of rabies in terrestrial carnivores because of the successful seeding oral vaccine for wild foxes.

Schistosomiasis

This disease is caused by the parasites of genus *Schistosoma* and is of two types. *Schistosoma mansoni* is responsible for intestinal schistosomiasis, while *Schistosoma haematobium* for the urogenital schistosomiasis. The parasites are found in tropical and subtropical Africa, South America and Asia. Schistosomiasis claims about 500,000 deaths each year.

The parasitic form which is infectious in humans is released into water by a mollusc, which thus serves as the intermediate host. The parasite infects humans, the definitive host, through the skin. Infections typically occur while swimming in fresh water or by simply standing or wading in stagnant water. About 2 – 4 days after infection, the microscopic worm leaves the skin and enters the bloodstream to reach the heart or lungs. In these tissues, it undergoes maturation and then again enters the bloodstream to reach the liver. Here it undergoes sexual differentiation and emerges as an adult male or female. The adult forms of *Schistosoma mansoni* remain in the veins of the intestine, while those of *Schistosoma haematobium* travel to the bladder. At this stage, the female begins to lay eggs continuously (300 eggs per day in case of *S. mansoni*). Some eggs get entrapped in

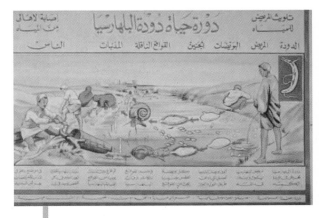

Life cycle of schistosomas.

the host tissue and give rise to granulomas, which are a characteristic pathological feature. Remaining eggs traverse the wall of the intestine or urinary bladder as the case may be (depending on the species), and are excreted from the human host through the feces or urine, respectively. The eggs hatch in fresh water and the highly motile larval forms emerge. The larvae infect the intermediate host, the mollusc, and multiply several times before being released into water once again, ready to infect a human host. This completes their life cycle.

Intestinal schistosomiasis gives diarrhea and colic, with blood in the stool if the illness becomes chronic. Urinary schistosomiasis is revealed the presence of blood in the urine. The body responds by synthesizing a kind of a capsule around the eggs, called granulomas, which can lead to serious complications. In case of *S. mansoni*, there can be complications such as obstruction of the bowels (preventing passage of gas and solid materials), vomiting of blood, liver cirrhosis, pulmonary fibrosis, or appendicitis, that develop rather rapidly. Complications from *S. haematobium* are ulcers, stones, renal failure, urinary fistulas (abnormal connections), increase in the volume of renal cavities due to excessive pressure, wart-like papillomas, or bladder cancer.

Effective treatments exist for these two major types of the disease.

Small pox

This a highly contagious disease caused by a the pox-virus that only infects humans. It is generally spread by the inhalation and skin contact. The incubation period is about fifteen days. The virus makes lesions in the pharynx that lead to airborne contamination. After a phase of fever, red patches appear on the skin. They soon transform into vesicles and pustules (loaded with the virus), especially on the face and palms. The pustules become crusty and disappear leaving behind indelible pock marks, characteristic of the disease. Small pox was fatal in 20 – 40% of the cases. If the patient recovers, there is life-long immunity to the disease.

The virus has been totally eradicated from our planet since 1977 due to a vaccine consisting of a closely related live virus (cow-pox virus). This vaccine was a great medical progress. The downside is that the population that is no longer vaccinated is highly susceptible to the disease and hence vulnerable to a small pox epidemic, for example caused by a malicious bioterrorist attack with the virus.

[Figure legend: Jenner vaccinating his son (Monteverde, 1878).]

Syphilis

Syphilis (sometimes called pox) is a sexually transmitted infection (STI) caused by spiral-shaped bacterium, *Treponema pallidum*. This bacterium is very vulnerable as it cannot survive in the outside environment. Transmission of the disease is through sexual contact. Syphilis occurs all around the world, including developed countries. About 500 new cases are reported in France per year.

Jenner vaccinating his son (Monteverde, 1878).

The disease occurs in 3 phases if care is not provided. Primary syphilis is characterized by the appearance of a painless, pink colored, ulcerated lesion at the tip of genital organs. This lesion, called chancre, appears about three weeks after intercourse. Symptoms of secondary syphilis are seen between one month to one year after the initial infection and include several lesions on the skin and mucous membranes, accompanied by enlargement of lymph nodes, fatigue, renal dysfunction and circulatory disorders. Tertiary syphilis may occur after a few months or a few years and is characterized by neurological disorders (neurosyphilis) leading to disabilities. Other organs may also be affected, such as the heart, liver, intestines, kidneys, eyes, etc. and sometimes psychiatric disorders are seen It is usually a cardiac arrest that causes death.

Penicillin is still the treatment of choice for syphilis.

It should be noted that a rare form of congenital syphilis may also occur, though rare in developed countries. It is transmitted from mother to child due to the passage of the treponemes across the placenta. The infection often leads to abortion or premature birth. In latter case, the baby is born with symptoms resembling those of secondary syphilis.

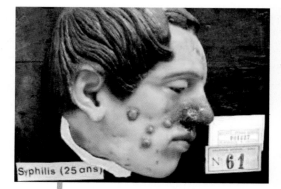

Syphilis (25 ans)

A wax figure showing a syphilis patient, Museum Orfila-Rouviere, Paris.

From the Dark Ages to Modern Times

There have been several celebrated personalities who have suffered from syphilis, to cite a few at random: Franz Shubert, Robert Schumann, Charles Baudelaire, Guy Maupassant, Vincent van Gogh, Lenin, Al Capone, Howard Hughes etc.

Tetanus

Tetanus is an infectious disease caused by a Gram positive bacterium, *Clostridium tetani*. It causes about 50,000 to 1 million deaths per year. While the mortality in around 1% in developed countries, it reaches 90% in African countries, especially when the infection is in neonates. Neonatal tetanus occurs when the umbilical cord is cut with material that is contaminated or in some cases due to application of a plaster made of mud!

The bacteria are found in soil, dust, plants and rusted objects. They live in the intestines of horses, cattle and humans. They are capable of forming resistant spores. The infection begins with contamination of a wound of broken skin with these bacteria or their spores. The bacteria are strict anaerobes, which means that they can proliferate only in the absence of oxygen. Conditions conducive to their growth are found in deep wounds as well as superficially such as in the case of frost bite, burns, ulcers or in cases of abortion. Needles used by drug addicts can also be a source of contamination.

When the *C. tetani* bacteria multiply, they secrete a toxin which reacts with a certain type of neuronal cells, leading to an increase in the motor activity of muscles that causes them contract or to go into spasms which are a characteristic of the disease. Muscle contractions may occur all over the body (generalized tetanus) and very painful. Death is usually due to respiratory failure.

Treatment is usually given in an intensive care unit and consists of muscle relaxants. Injection of anti- toxin sera inhibits its action. Treatment with antibiotics stops proliferation of the bacteria. The mandatory anti-tetanus vaccine is very effective.

Tuberculosis

This is an infectious disease that is currently showing a renewed upward trend with 10 million new cases recoded every year worldwide. It is a particularly serious public health issue in developing countries like Asia, South America and Africa. Risk factors include alcoholism, malnutrition, AIDS and chemotherapeutic treatment. Tuberculosis mostly affects the underprivileged social classes such as the homeless, immigrants, prisoners or drug addicts.

The causative agent is the bacterium, *Mycobacterium tuberculosis*, also called Koch's bacillus, that cannot be classified by the Gram staining method. Humans are the only reservoir of the bacterium. The disease spreads through droplets of respiratory secretions from infected individuals. Transmission is by inhalation. On the first contact with the

A poster in French to popularize the BCG vaccine. The message conveyed by the poster is: "They are not worried about Tuberculosis anymore – they are vaccinated".

bacteria, the lungs respond by trying to trap them in an immune reaction aimed at their destruction. This is the primo-infection. In 9 cases out of 10, the infection stops here as the patient recovers and also develops life-long immunity. In other cases, the bacteria can invade any and all organs (lungs, kidneys, bones, brain) and symptoms of tuberculosis appear. Typical symptoms include weight loss, loss of energy, reduced appetite, fever, cough with sputum tainted with blood. Immunodeficient persons contract the disease easily. This is why 10 – 15% of HIV-AIDS patients die from tuberculosis. These two diseases have evolved in parallel, particularly in Africa.

Diagnosis involves a skin test (intradermal reaction). Treating the primo infection is very important in order to prevent the disease from progressing. Once the disease is triggered, the required antibiotic treatment is very long, lasting 6 – 9 months. Currently tuberculosis treatment has become difficult because the bacteria have become increasingly resistant to different antibiotics.

The vaccine comprises of the strain called BCG for Bacillus of Calmette and Guerin. It is not 100% protective, and is no longer compulsory in France but strongly recommended for children at high risk.

Tularemia

This is an infectious disease caused by a Gram negative bacterium, *Franciscella tularensis*. Small rodents and hare are susceptible to this disease, while, some other animals may serve as vectors without being sick, for example, boars, cats, dogs, foxes and biting insects like ticks, horseflies and mosquitoes. The bacteria are found in the feces of infected animals. In France, there are about 50 cases reported per year. Rarely, the bacteria are able to pass through healthy skin. Handling of an infected hare by a hunter can result in sufficient skin contact for the bacteria to be transmitted.

Hare, a vector of the tularemia.

The disease begins with sudden high fever, chills, intense fatigue, sometimes pain in the joints and muscles, headache and occasionally nausea and vomiting. In the case of skin contact, lymph nodes swell and fistulas appear. In the case of ocular contact, there is conjunctivitis while pulmonary tularemia is characterized by respiratory distress. Sometimes, a digestive form of the disease occurs with diarrhea. Some forms can lead to death after a coma if the bacteria get into the bloodstream (septicemia).

Treatment is with antibiotics.

The best method of prevention is to avoid all contact with sick animals. In regions suspected to be have the disease, drinking water should be treated and meat of wild animals should be well cooked.

These bacteria have been proposed as a biological weapon in the form of aerosols.

Typhoid fever

Typhoid fever is an infectious disease caused from ingestion of food or water contaminated with the Gram negative bacterium, *Salmonella* Typhi. It affects about 30 million people per year, with Asia and Africa being most affected, as its incidence in developed countries is rare. The bacteria are released with the feces of typhoid patients and under conditions of poor hygiene, contaminate food and water causing a spread of the disease. After a transient diarrhea, the bacteria that have proliferated in the lymph glands, begin to spread through the blood. The patient then has fever, enlarged spleen, abdominal pain and intense fatigue. There are signs of prostration (the "tuphos") and death can occur from hemorrhagic complications in the digestive system in 30% of the untreated cases. Antibiotic therapy can stop the illness. A highly effective vaccine called Typhim Vi is available.

Typhoid Mary.

There are also "healthy carriers", that is persons who are infected with the bacterium, but do not show any symptoms. Nevertheless, they release the bacteria in their feces and can thus contaminate their environment, unknowingly. The first known example of a healthy carrier was an Irish-American cook called Mary Mallon (1869 – 1938), alias "Typhoid Mary", who infected 32 people, of which 2 died. She refused to take the treatment prescribed and had to be quarantined.

Typhus

This is a disease transmitted by an unusual type of bacteria called *Rickettsia prowazekii*, that can survive only inside host cells. It is fatal in 10-30% of cases. This disease is still present in some countries in Africa, Latin America and Asia. It is transmitted to humans by body lice and is therefore favored by poverty, in densely populated areas, during wars, people movements, etc. and especially in cold seasons.

Body louse, Pediculus humanus, a vector of the Rickettsia prowazekii, site of Timone.

Scene of delousing. Detail from a painting, Heart of a farm, by Jan Sieberechts, 1662. Museum of Fine Arts, Bruxelles.

After an incubation period of 10 – 14 days, the disease begins with a sudden fever, chills, headache, pain in joints and muscles. There may be skin eruptions and nervous system disorders such as confusion. In the severe form of the disease, there may be pneumonia, leading to death.

The bacteria are considered a potential biological weapon.

Yellow fever

Yellow fever is an infectious disease caused by a virus and transmitted by mosquitoes. The name comes from the fact that some patients turn yellow from jaundice that they may get during the illness. The disease is seen in tropical Africa and Americas. There are an estimated 200,000 infections and 30,000 deaths per year. The incidence of the disease is increasing steadily and is a serious public health issue.

Mosquito, Aedes.

The disease affects humans and monkeys, but the reservoir for the virus are mosquitoes of the species, *Aedes egypti* and *Aedes haemagogus*, that transmit the virus by their bite. Mosquito eggs that are infected with the virus in one wet season can release adult mosquitoes carrying the virus in the next wet season, as the eggs can survive a period of dryness. The first phase or the "acute phase" of yellow fever begins three to six days after a bite of an infected mosquito. This consists of fever, muscular pain, headache, chills, nausea and vomiting. About 3 – 4 days later the symptoms disappear and the patients feel better in most cases. In a small number of patients (15%), the fever reappears within 24 hours and the "toxic" phase begins. These patients show signs of jaundice, abdominal pain, nausea and vomiting. There may be blood in the vomit and stools and bleeding from the nose, eyes and mouth. Renal functions are impaired. Approximately half of these patients die in 10 days while the others recover without further consequences.

There is no specific treatment for yellow fever. Treatment is aimed at bringing the fever down and antibiotics are given against possible secondary infections. The use of insecticide-treated mosquito-nets is recommended. Mosquito control measures have been shown to reduce the incidence of the disease. A vaccine that is effective for 10 years is available.

Areas at risk of Yellow Fever transmission

Worldwide distribution of yellow fever.

References

Altman L.J., *Plague and pestilence. A history of infectious disease.* Enslow Publishers, Berkeley Heights, NJ, USA, 1998.

Bollet A.J., *Plagues and poxes. The impact of human history on epidemic disease.* Demos, New York, 2004.

Cartwright F.F., Biddiss M., *Disease and history*, 2nd edition, Sutton Publishing Limited, Phoenix Mill, UK, 2000.

Chastel C., *Virus émergents. Vers de nouvelles pandémies ?* ADAPT-SNES Éditions, Paris, 2006.

Christopher G.W., Cieslak T.J., Pavlin J.A., Eitzen E.M., *Biological warfare. A historical perspective.* JAMA, 278, 412-417.

Clarke R., *La guerre biologique est-elle pour demain ?* Fayard, Paris, 1972. Colnat A., *Les épidémies et l'histoire.* Éditions Hippocrate, Paris, 1937.

Derbes V.J., *De Mussis and the great plague of 1348.* JAMA, 1966, 196, 179-182. Geissler E., van Courtland Moon J.E., *Biological and toxin weapons: research, development and use from the Middles Ages to 1945.* Sipri, Oxford University Press, New York, 1999.

Hansen W., Freney J., *Le charbon : maladie d'hier, arme biologique d'aujourd'hui.* Pour la Science, 2001, 290, 8-15.

Henderson D.A., *Bioterrorism as a public health threat.* Emerg. Infect. Dis. 1998, 4, 488-492.

Klein D., *La guerre microbienne.* Thèse de médecine, faculté de médecine et de pharmacie de Lyon, 1935-1936.

Klietmann W.F., Ruoff K.L., *Bioterrorism: implication for the clinical microbiologist.* Clin. Microbiol. Rev. 2001, 14, 364-381.

Lesho E., Dorsay D., Bunner D., *Feces, dead horses and fleas. Evolution of the hostile use of biological agents.* WJM, 1998, 168, 512-516.

Mayor A., *Greek fire, poison arrows, and scorpion bombs.* Overlook Duckworth, Woodstock & New York, 2003.

Miller J., Engleberg S., Broad W., *Germes.* Fayard, Paris, 2002.

Mollaret H.-H. *Bref historique de la guerre bactériologique.* Med. Mal. Infect.1985, 7, 402-406.

Mollaret H.-H. *Contribution à l'histoire du crime bactérien*. Med. Mal. Infect. 1987, 2, 56-57.

Mollaret H.H., *La mort de Heydrich : un cas très spécial de botulisme*. Med. Mal. Infect. 1986, 8/9 493-495.

Oldstone M.B., *Viruses, plagues & history*. Oxford, New York, 1998.

Perrut J.-J. *Risques et menaces biologiques*. Éditions du Paradis, Cornebarrieu, France, 2003.

Preston R., *Les nouveaux fléaux*. Plon, Paris, 2003.

Prinzing F., *Epidemics resulting from wars*. Oxford – Clarendon Press, Humphrey Milford, 1916.

Riche D., *La guerre chimique et biologique*. Pierre Belfond Éd., 1983.

Ruffié J., Sournia J.-C., *Les épidémies dans l'histoire de l'homme*. Flammarion, Paris, 1993.

Sherman I.W., *Twelves diseases that changed our world*. ASM Press, Washington D.C., 2007.

Simmon J.D., *Biological terrorism, preparing to meet the threat*. JAMA, 1997, 428-430. Warry J., Warfare in the classical world. University of Okhlahoma Press, Norman, 1995.

Werner A., Werner H., Goetschel N., *Les épidémies, un sursis permanent*. Éditions Atlande.

Zinsser H., *Rats, lice and history*. Little, Brown and co., Boston, 1963.

Copyright
for "Microbes at War"

Acknowledgements
for figures and illustrations

Sources

Agence Bridgeman Giraudon : pp. 62 and 69

Agence RMN Musées Nationaux : pp. 14, 16, 18, 37,52, 58 and 80

Agence AP / Sipa Press : pp. 94, 98 and 99

Musée du Capitole, Rome, Italie : pp. 23 and 26

Sources

Figure p. 12 :
fr.wikipedia.org/wiki/Saturne_dévorant_un_de_ses_enfants

Figure p. 13 : de.wikipedia.org/wiki/Julius_Schnorr_von_Carolsfeld

Figure p. 14 : commons.wikipedia.org/wiki/Image:Folio_29r_-_The_Ark_of_God_Carried_into_the_Temple.jpg

Figure p. 15 : karenswhimsy.com/.../images/assyrians-3.jpg (sennacherib)

Figure p. 17 : environnement.ecoles.free.fr (Un ange détruit l'armée de Sennachérib, Gustave Doré)

Figure p. 17 : fr.wikipedia.org/wiki/Pericles

Figure p. 19 : en.wikipedia.org/wiki/Alexander_the_Great (Alexander the Great on his horse Bucephalus (Pompei))(sarcophage Istambul)

Figure p. 21 : fr.wikipedia.org/wiki/Xénophon

Figure p. 22 : www.histoire-fr.com/images/Brennus

Figure p. 22 : fr.wikipedia.org/wiki/Alcibiade

Figure p. 25 : fr.wikipedia.org/wiki/Huns

Figure p. 25 : en.wikipedia.org/wiki/Raphael_Rooms

Figure p. 26 : fr.wikipedia.org/wiki/Marc_Aurèle

Figure p. 27 : en.wikipedia.org/wiki/Cyprian

Figure p. 28 : fr.wikipedia.org/wiki/Antioche

Figure p. 31 : fr.wikipedia.org/wiki/Hannibal_Barca

Figure p. 32 : fr.wikipedia.org/wiki/Frédéric_Barberousse

Figure p. 33 : en.wikipedia.org/wiki/Battle_of_Vicksburg

Figure p. 34 : www.castlemaniac.com/.../siege-chateau.gif

Figure p. 34 : mirimen.com/co_beo/Sigizmund-Koribut-2EFA.html

Figure p. 38 : en.wikipedia.org/wiki/Justinian_I

Figure p. 40 : fr.wikipedia.org/wiki/Peste_noire

Figure p. 41 : fr.wikipedia.org/wiki/Pieter_Bruegel_l'Ancien

Figure p. 44 : en.wikipedia.org/wiki/Hans_Karl_von_Diebitsch

Figure p. 45 : fr.wikipedia.org/wiki/Roméo_et_Juliette

Figure p. 46 : en.wikipedia.org/wiki/Charles_XII_of_Sweden

Figure p. 48 :
fr.wikipedia.org/wiki/Déclin_de_l'Empire_romain_d'Occident

Figure p. 48 : fr.wikipedia.org/wiki/Esclavage

Figure p. 49 : fr.wikipedia.org/wiki/Toussaint_Louverture

Figure p. 50 : fr.wikipedia.org/wiki/Louisiane

Figure p. 51 : fr.wikipedia.org/wiki/Christophe_Colomb

Figure p. 53 : fr.wikipedia.org/wiki/Alexandre_VI

Figure p. 53 : fr.wikipedia.org/wiki/François_Ier_de_France

Figure p. 54 : fr.wikipedia.org/wiki/Henri_VIII_d'Angleterre

Figure p. 54 : en.wikipedia.org/wiki/Ivan_IV_of_Russia

Figure p. 56 : fr.wikipedia.org/wiki/Francisco_Pizarro

Figure p. 57 : en.wikipedia.org/wiki/Jeffrey_Amherst

Figure p. 77 : fr.wikipedia.org/wiki/Guerre_de_Trente_Ans

Figure p. 63 : fr.wikipedia.org/wiki/Louis_IX_de_France (Damiette)

Figure p. 64 : fr.wikipedia.org/wiki/Bataille_d'Azincourt

Figure p. 65 : fr.wikipedia.org/wiki/Bataille_de_Valmy

Figure p. 68 : fr.wikipedia.org/wiki/Guerre_des_Boers

Figure p. 70 : fr.wikipedia.org/wiki/Grippe_espagnole

Figure p. 71 : en.wikipedia.org/wiki/American_Expeditionary_Force

Figure p. 71 : fr.wikipedia.org/wiki/Woodrow_Wilson

Figure p. 72 : fr.wikipedia.org/wiki/Erich_Ludendorff

Figure p. 72 : fr.wikipedia.org/wiki/Georges_Clemenceau

Figure p. 73 : en.wikipedia.org/wiki/Irish_Potato_Famine_(1845-1849)

Figure p. 74 : fr.wikipedia.org/wiki/John_Fitzgerald_Kennedy

Figure p. 76 : fr.wikipedia.org/wiki/Tétanos

Figure p. 76 : fr.wikipedia.org/wiki/Pythagore

Figure p. 85 : fr.wikipedia.org/wiki/Reinhard_Heydrich

Figure p. 88 : fr.wikipedia.org/wiki/Première_Guerre_mondiale

Figure p. 88 : fr.wikipedia.org/wiki/Bataille_de_Stalingrad

Figure p. 91 : fr.wikipedia.org/wiki/Shiro_Ishii

Figure p. 94 : fr.wikipedia.org/wiki/Saddam_Hussein

Index

Charles VIII : pp. 51, 52, 54, 66
cholera : pp. 11, 68-70, 74, 83, 91, 94, 106, 109
Claviceps purpurea : p. 108

D

Dagobert : p. 40
Damas
dengue : pp. 7, 18, 106, 107
dysentery, amoebic : pp. 107, 108
dysentery, bacillary or shigellosis : pp. 107, 108

E

English sweating sickness : pp. 66, 67
Entamoeba histolytica : pp. 107, 108
ergot of rye : pp. 18, 26, 108
ergotism : pp. 18, 19, 26, 108

F

flagellants : p. 43
Fort Detrick : p. 91
fowl cholera : p. 83, 109
Francis I : pp. 53, 59, 60
Francisella tularensis : pp. 88, 89
fungus : pp. 26, 73, 94, 103, 108, 115

G

gas gangrene : pp. 30, 109
Geneva protocol : pp. 70, 95, 96
glanders : pp. 87, 88, 110
Gram staining : pp. 102, 105, 111, 119
Gruinard islands : p. 90

H

Hannibal : pp. 31, 32
Hemophilus influenzae
Henry Girard : p. 84
Henry VIII : pp. 53, 54, 66
Hercules and Deianeira : p. 80
Heydrich : pp. 85, 86, 126
Huns : pp. 24, 25

I

influenza, bird flu : p. 110
influenza, Spanish flu : pp. 70, 71, 87, 110
Iraq : p. 94
Iroquois : p. 33
Island of Renewal : p. 92
Ivan the Terrible : p. 54

J

Japan : pp. 55, 90-92, 98
Jericho : pp. 13, 14
Jerusalem : pp. 15, 17
jews : pp. 17, 42, 43, 78
Justinian : pp. 38, 39

From the Dark Ages to Modern Times

From the Dark Ages to Modern Times

T

U

U

W

Y

Printed in Spain:
Graficas LIZARRA S.L.
31132 Villatuerta - Navarra - Spain
October 2011